Lecture Notes in Computer Scien

T0230302

Commenced Publication in 1973
Founding and Former Series Editors:
Gerhard Goos, Juris Hartmanis, and Jan van Leeuwen

Miroslaw Malek   Edgar Nett
Neeraj Suri (Eds.)

Service Availability

Second International Service Availability Symposium, ISAS 2005
Berlin, Germany, April 25 – 26, 2005
Revised Selected Papers

 Springer

Volume Editors

Miroslaw Malek
Humboldt-Universität zu Berlin
Institut für Informatik
Rechnerorganisation und Kommunikation
Rudower Chaussee 25, 12489 Berlin, Germany
E-mail: malek@informatik.hu-berlin.de

Edgar Nett
Otto-von-Guericke-Universität Magdeburg
Institut für Verteilte Systeme
Universitätsplatz 2, 39106 Magdeburg, Germany
E-mail: nett@ivs.cs.uni-magdeburg.de

Neeraj Suri
Technical University Darmstadt
Department of Computer Science
Dependable Embedded Systems and Software
Hochschulstr. 10, 64289 Darmstadt, Germany
E-mail: suri@informatik.tu-darmstadt.de

Library of Congress Control Number: 2005932545

CR Subject Classification (1998): C.2, H.4, H.3, I.2.11, D.2, H.5, K.4.4, K.6

ISSN   0302-9743
ISBN-10   3-540-29103-2 Springer Berlin Heidelberg New York
ISBN-13   978-3-540-29103-9 Springer Berlin Heidelberg New York

Springer is a part of Springer Science+Business Media

springeronline.com

© Springer-Verlag Berlin Heidelberg 2005
Printed in Germany

Typesetting: Camera-ready by author, data conversion by Scientific Publishing Services, Chennai, India
Printed on acid-free paper   SPIN: 11560333   06/3142   5 4 3 2 1 0

General and Program Chairs' Message

The 2nd International Service Availability Symposium (ISAS 2005) provided a unique forum for academia and industry researchers who focus not only on developing next generation solutions but also on standards for today's market. Given the pervasive interweaving of computing devices, increasingly it is "services" rather than "systems" that warrant our attention. As services emerge as the primary vehicle for information acquisition, processing and delivery, the demands for dependability become of primary concern. Needless to add, the expectations from users' with respect to trust and reliance of such systems will only continue to grow.

As computers already pervade almost all walks of our lives, significantly increased interest in dependable computing should not be a surprise as the industry leaders and main computer companies are searching for innovative ways of enhancing the dependability of systems that are increasingly more complex and networked. With the paradigm shift where "everything" may become a service, it is not an option but an imperative to address the questions of service availability. From humble beginnings of dealing with types and formats, later with tasks and processes, then with objects and components, we have arrived to service and peer-to-peer computing. Over 8.5 billion processors are produced each year and 98.5% end up in geographically distributed and interconnected embedded systems. The challenge is to design services and systems that are highly available, reliable and secure. As the number of 7×24 applications continuously increases this is an ambitious challenge that will have to be met. Service availability cannot be compromised. It will have to be delivered as the economic and social impacts of unreliable, incorrect services might range from minor inconveniences to losses of human lives and unpredictable costs.

This year's ISAS represented an excellent mix of academic and industrial contributions as well as participation.

The eight sessions featured truly distinguished academics and industrial leaders as well as some new researchers in the field. We had an outstanding Keynote Speaker Prof. Hermann Kopetz from TU Vienna who is a pioneer in the field of dependable real-time computing, actively contributing to the field for almost 30 years. A distinguished panel featuring representatives from academia and industry, two invited sessions, and regular papers that were subject to a rigorous review process constituted the overall ISAS program. Each paper was reviewed by at least three Program Committee members. We would wholeheartedly like to thank our PC members for their guidance and diligent reviewing. Our thanks go to Prof. Edgar Nett, Nikola Milanovic and Christine Henze for editing the proceedings. Nikola and Christine also helped together with Sabine Becker and Steffen Tschirpke of Humboldt University Berlin, Susan Morgner and Dr. Christine Titel from Congressa GmbH the organization and we do appreciate it very much! Last but not least we would like to thank Manfred Reitenspieß who has been the guiding force behind ISAS and the Service Availability Forum.

I hope that the attendees enjoyed the final program, enjoyed the presentations, got involved in the discussions, struck up new friendships, and got inspiration for contributions to the next year's symposium which will be hosted by Kimmo Raatikainen, University of Helsinki and Francis Tam of Nokia in Helsinki during May 15–16, 2006.

Miroslaw Malek
Humboldt Universität Berlin
Institut für Informatik
malek@informatik.hu-berlin.de
ISAS 2005 General Chair

Neeraj Suri
Technische Universität Darmstadt
Institut fuï Informatik
suri@informatik.tu-darmstadt.de
ISAS 2005 Program Chair

Table of Contents

TTA Supported Service Availability

H. Kopetz

Institut für Technische Informatik TU Wien,
A 1040 Wien, Treitlstrasse 3
hk@vmars.tuwien.ac.at

Abstract. The Time-Triggered Architecture (TTA) is a distributed architecture for high-dependability real-time applications. In this paper the mechanisms that guarantee a high availability of TTA services are presented. The paper starts with a deliberation on the fault-hypothesis of the TTA and discusses the partitioning of a TTA system into independent fault-containment regions, their failure modes and their failure frequencies. In the second part the structure of the TTA is explained and the mechanisms that handle the specified faults are outlined. The role of the TTA-inherent sparse time base for the consistent ordering of messages and the solution of the simultaneity problem is explained. Finally, the third part speculates on the vision of a highly integrated TTA-giga-chip that acts as a self-contained TTA node and could be implemented on a single silicon die.

1 Introduction

The Time-Triggered Architecture (TTA) [1] is an *integrated distributed* computer architecture, designed to provide a continuous *timely* services with an MBTF of better than 10^9 hours in the presence of component failures, *provided that* the occurrences of component failures are in agreement with the stated fault hypothesis. The TTA is intended for applications that require utmost availability even in the presence of a fault in any of its components: examples of such applications are the control of a nuclear power plant, the flight control system of an airplane or a computer-based brake control system within an automobile that does not contain a mechanical backup.

Such a high reliability can only be achieved by the provision of redundancy in the hardware, since the observed component (chip) failure rates are orders of magnitudes lower [2] than the desired system reliability. Every redundancy scheme is based on a number of assumptions--*the fault hypothesis*--about the types and frequency of faults that the system is supposed to handle. In case that all fault-handling mechanisms are perfect and cover all scenarios that are listed in the fault-hypothesis, the probability of system failure is reduced to the *assumption coverage*[3], i.e., the probability that the assumptions made in the fault hypothesis are met by reality. The fault hypothesis of any fault-tolerant system is a critical document in the design process. The fault hypothesis of the TTA is discussed in more detail in Section two.

One common technique to implement fault-masking by redundancy is called *triple-modular redundancy* (TMR). In a TMR system fault-tolerant units (FTUs) are formed by placing three synchronized deterministic replicas of every critical component into a new distributed unit--the FTU. An incoming message is distributed

M. Malek, E. Nette, and N. Suri (Eds.): ISAS 2005, LNCS 3694, pp. 1 – 14, 2005.

to all three units of the FTU and the result message (and the internal state) is output-
ted to a voter that makes a majority decision based on at least two identical results. If
one of the components of FTU produces no result or a result that is different from the
result of the other two components, this component is considered to have failed.
TMR structures will only succeed if the redundant components fail *independently*,
i.e. if there is no correlation between the failures of components that form a fault-
tolerant unit. Correlated failures can occur because of external causes (a single external event, e.g., a lightning stroke, causes the failure of more than one component) or
by error propagation, i.e. an erroneous component sends a faulty message to an up to
that instant correctly operating component and thus corrupts the internal state of this
component. The issues of fault isolation and error propagation in the TTA are covered in Sections three and four.

Finally in Section five and six we speculate about the future hardware implementation of the TTA. Considering the tremendous advances in the field of semiconductor
technology, which is expected to give us billion-transistor giga chips (system-on-a-
chip: SoC) by the end of this decade, we outline the structure of a generic TTA-
SoC that can be used in many different application domains.

2   Fault Hypothesis of the TTA

In the following paragraphs we discuss the fault hypothesis of the TTA with respect
to hardware faults. We assume that the hardware design and the basic fault-handling
mechanisms are free of design faults.

2.1   Fault-Containment Regions

The first step in the specification of a fault-hypothesis is concerned with the establishment of a the fault-containment regions (FCR), i.e. the *units of failure*. An FCR
is a subsystem that is considered to fail independently from any other FCR. If we
must tolerate the physical destruction of a hardware component (e.g., in an accident),
then different FCRs must be in different physical locations, i.e. the computer system
must be distributed. In the TTA we assume that every node of the distributed system
forms an FCR.

2.2   Failure Modes

In the next step we must specify the *critical failure modes* of FCRs. Any restriction
of the tolerated failure modes must be considered as an *additional assumption* that
has a negative effect on the assumption coverage. In the optimal case no restriction
of the failure modes are made, i.e. a failing component can manifest an arbitrary
behavior.

We consider a failure mode of an FCR as *critical*, if it impacts the remaining correct nodes of the distributed system in such a way that the functionality or the consistency of the distributed computing base among the nodes that are outside the affected
FCR is lost. We focus on a single fault during a fault-recovery interval Δd. After the
recovery interval Δd the architecture has recovered from the consequences of this
fault and can tolerate a further fault (provided enough resources remain operational).

We define a set of nodes as Δd-*consistent* if Δd time units after the occurrence of failure all remaining correct nodes have the same view about this failure event.

In the TTA we have identified the following critical failure modes of an FCR that must be addressed at the level of the architecture:

(i)   Crash/Omission (CO) failures
(ii)   Babbling idiot failures
(iii)   Slightly-off-specification (SOS) failures
(iv)   Masquerading failures
(v)   Massive transient disturbances

In the analysis of failure mode (i) to (iv) we assume that a fault impacts a single FCR only. Failure mode (v), special case that affects more than one FCR, is explained at the end of this Section.

Crash/Omission Failures: A widely accepted fault-model in a distributed system assumes a fault that manifests itself as either a crash failure of a node or an omission failure of the communication channel (CO failure). CO failures are the most common failures in distributed system--close to 99% of the failures are of the CO type[4]. According to this fault model a node either operates correctly or crashes. The communication system either transports a message correctly, produces a detectably corrupted message, or fails to transport a message. Most of the available communication protocols, such as for example TCP/IP, are designed to detect and, if possible, to correct CO failures. The consequence of a CO failures is a loss of consistency of the distributed computing base. In a point-to-point communication system an acknowledgement service is provided to detect CO failures. In multi-cast communication system, such as the TTA, a membership service can is available to detect and identify CO failures. Another mechanism for CO failure detection is the acknowledgement mechanism of the CAN protocol[5].

In a multicast environment it is important to distinguish between an omission failure at the sender and an omission failure at one of the receivers. If the sender learns promptly about a local omission failure it can often undo the state-change assumed to have taken place by rolling back to the state before the send operation. In case of an omission failure at one of the receivers, such a rollback is not possible.

Prompt CO failure detection and diagnosis at the architecture level is important in order to inform the application that consistency has been lost, and which unit is responsible for the loss of consistency. The application can then decide what corrective action must be taken.

Babbling idiot failures: A babbling idiot failure of an FCR occurs, if the FCR starts sending *untimely messages*. In a multicast time-triggered communication topology that contains a broadcast channel such a babbling FCR can interfere with the communication of the correct nodes. If an FCR exhibits permanent babbling-idiot failures on both channels (this is in principle possible, since both channels are in the same FCR) any further communication among the correct nodes becomes impossible. The TTA detects and handles babbling-idiot failures of FCRs by the guardians in the communication system. The guardian will only open the sending channels during the *a priori* known time-interval that has been allocated to a node.

Slightly-off-Specification (SOS) Failures: Slightly-off-specification failures are an important special case of Byzantine failures. They can occur at the interface between the analog and the digital world. Assume the situation as depicted in Figure 1. The specification requires that every correct node must accept analog input signals if they are within a specified receive window of a parameter (e.g., timing, frequency, or voltage). Every individual node will have a wider actual receive window than the one specified in order to ensure that even if there are slight variations in manufacturing it can accept all correct input signals as required by the specification. These actual receive windows will be slightly different for the individual nodes, as shown in Fig 4. If an erroneous FCR produces an output signal (in time or value) slightly outside the specified window, some nodes will correctly receive this signal, while others might fail to receive this signal. Such a scenario will result in an inconsistent state of the distributed system.

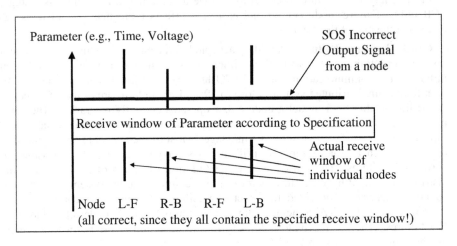

Fig. 1. Slightly off Specification (SOS) failure

Example: Consider a brake-by-wire system where four receiving nodes are at the four wheels of a car (L-F: left front, R-B: right back, R-F: right front, L-B: Left back). In this example an SOS output failure of the "brake master" will cause confusion in the distributed system. According to this example, the *L-Front* and the *L-Back* node will accept the SOS message, while the *R-Back* and *R-Front* node will discard this message. In a brake-by-wire system, such an inconsistency can become safety relevant.

In the TTA we must address the following three types of SOS failures:

(i)   SOS value failure
(ii) SOS coding failures
(iii) SOS send-instant failures.

An *SOS value failure* occurs, if the signal level of the outgoing message is SOS faulty. Some receivers may still correctly decode such an SOS faulty signal, while others may not be able to decode this signal. Since both outgoing channels of an FCR depend on the same power supply, the probability that SOS value failures on both channels are correlated.

An *SOS coding failure* occurs, if the bit stream from the sender is at the border of the coding specification, e.g., the frequency is SOS faulty. Since both channels are driven by the same oscillator the probabilities for the occurrence of an SOS coding failure on both channels are not independent.

An *SOS send-instant failure* occurs, if the send-instant of a message transmission (see Section 3.1) is SOS faulty. A message that is SOS send-instant faulty may be accepted by some nodes and rejected by others. Again, SOS send-instant failures on the two channels are correlated.

Masquerading Failures: A masquerading failure occurs if an erroneous node assumes the identity of another node and causes harm to the system. Systems that rely on names stored in a message to identify the transported message and the information contained therein are vulnerable to masquerading failures. It opens the possibility that a single faulty node can masquerade other nodes, without the receivers having a chance to detect the fault. For example, if a bit in the name of a message to-be-sent that is stored in the sending node is incorrect, this message could, after arrival at its destination, overwrite correct messages at correct receivers. This problem is discussed at some length in the safety-critical SafeBus protocol [6] p.36: *Any protocol that includes a destination memory address in a message is a space-partitioning problem.*

Massive Transient Disturbances: Another important fault class in a distributed embedded system, particularly in the automotive domain, is concerned with massive transient disturbances, e.g., those caused by electromagnetic emission (EMI). A massive transient disturbance can cause the temporary loss of communication among otherwise correct nodes that reside in different FCRs or cause state-corruptions within more than one node. Based on available failure data [2] it is reasonable to assume that the multiple correlated faults produced by a massive transient disturbance are transient, i.e. that the hardware is not faulted by the massive transient disturbance. In such a situation the architecture can provide the service of prompt error detection in order that the nodes may take some local corrective action until the transient disturbance has disappeared and the communication service and the consistency of the nodes is reestablished by a fast restart. For example, [7] report that in an automotive environment a temporary loss of communication of up to 50 msec can be tolerated by freezing the actuators in the positions that were taken before the onset of the transient disturbance. The probability of occurrence of transient disturbances must be reduced by proper quality engineering, e.g., by shielding the cables or installing fiber optics instead of copper. In a safety-critical distributed system massive transient disturbances must be rare events. From the point of view of the communication system, fast detection of a transient disturbance and fast recovery after the transient has disappeared are important.

2.3 Frequency of Faults

The assumptions about the frequency of fault occurrence are depicted in Table 1. We distinguish between transient failures and permanent failures as well as between fail-silent failures and Byzantine failures.

Table 1. Assumed failure rates

Type of Failure	Failure Rate	Source
permanent fail silent	< 100 FIT (MTTF > 1 000 000 hours)	Field data from the auto-motive industry[2]
transient fail silent	< 100 000 FIT (MTTF > 1000 hours)	SEUs caused by neutrons[8]
permanent Byzantine	< 2 FIT (MTTF 50 000 000 hours)	Fault injection experiments[4]
transient Byzantine	< 2 000 FIT (MTTF > 50 000 hours)	Fault injection experiments[4]

Whereas the data in line one--permanent failures--is derived from extensive field data, the assumptions of line two, three and four are not as well supported by experimental data and field evidence. In particular it is very difficult to find a good estimate for the transient failure rates, because these failure are very dependent upon the environmental conditions (e.g., geometry of the setup determines the susceptibility with respect to EMI, geographical position and altitude determines the rate of SEUs etc..) of the unit under observation. The failure rates of Table 1 are our best estimates and are used in our reliability models to calculate the service availability of the TTA.

3   Structure of the TTA

The time-triggered architecture (TTA) is a distributed architecture for the implementation of hard real-time applications. It consists of a set of nodes interconnected by a TDMA (time-division multiple access) based real-time communication system. The TTA provides the following services to the application at the architecture level

(i)   a consistent distributed computing platform with prompt error detection if consistency is lost by a failure that can be detected at the architecture level.
(ii)   a fault-tolerant global sparse time base of known precision at all nodes
(iii) mechanisms for the precise operational specification of the interfaces among the nodes in the domains of time and value. These interfaces are called "temporal firewalls".
(iv)   error containment such that arbitrary node failures can be tolerated
(v)   mechanisms that support the transparent implementation of fault-tolerance.

In the following section we will discuss two essential characteristics of the TTA, the TTA view of time and state.

3.1   Global Sparse Time

For most applications, a model of time based on Newtonian physics is adequate. In this model, real time progresses along a *dense* timeline, consisting of an infinite set of *instants*, from the past to the future. A *duration (or interval)* is a section of the timeline, delimited by two instants. A happening that occurs at an instant (i.e., a cut of the timeline) is called an *event*. An *observation* of the state of the world at an instant is thus an event. The *time-stamp* of an event is established by assigning the state of the local clock of the observer to the event immediately after the event occurrence. Due to the impossibility of synchronizing clocks perfectly and the denseness property of real time, there is always the possibility of the following sequence of events occurring: clock in component *j* ticks, event *e* occurs, clock in component *k* ticks. In such a situation, the single event *e* is time-stamped by the two clocks *j* and *k* with a difference of one tick. The finite precision of the global time-base and the digitalization of the time make it impossible in a distributed system to order events consistently on the basis of their global time-stamps based on a *dense* time. This problem is solved by the introduction of a *sparse time base* in the TTA. In the sparse-time model the continuum of time is partitioned into an infinite sequence of alternating durations of *activity* and *silence* as shown in Figure 2. The activity intervals form a synchronized system-wide *action lattice*. From the point of view of temporal ordering, all events that occur within a duration of activity of the action lattice are considered to happen *at the same time*. Events that happen in the distributed system at different components at the same global clock-tick are thus considered *simultaneous*. Events that happen during different durations of activity (at different points of the action lattice) and are separated by the required interval of silence (the duration of this silence interval depends among others, on the precision of the clock synchronization [9]) can be temporally ordered on the basis of their global timestamps. The architecture must make sure that significant events, such as the sending of a message, or the observation of the environment, occur only during an interval of activity of the action lattice. The time-stamps of events that are based on a sparse time base can be mapped on the set of positive integers. It is then possible to establish the temporal order of events by integer arithmetic.

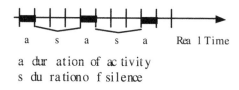

a dur ation of ac tivity
s du rationo f silence

Fig. 2. Sparse time base

The timestamps of events that are outside the control of the distributed computer system (and therefore happen on a dense timeline) must be assigned to an agreed lattice point of the action lattice by an *agreement protocol*. Agreement protocols are also needed to come to a system-wide consistent view of analogue values that are digitized by more than one analogue-to-digital converter.

3.2 Distributed State

In *abstract system theory*, the notion of *state* is introduced in order to separate the *past* from the *future* [10] p.45:

"The state enables the determination of a future output solely on the basis of the future input and the state the system is in. In other word, the state enables a "decoupling" of the past from the present and future. The state embodies all past history of a system. Knowing the state "supplants" knowledge of the past. Apparently, for this role to be meaningful, the notion of past and future must be relevant for the system considered."

Taking this view it follows that the notions of state and time are inseparable. If an event that updates the state cannot be said to coincide with a well-defined tick of a global clock on a sparse time-base, then the notion of a system-wide state becomes diffuse. It is not known whether the state of the system at a given clock tick includes this event or not. The *sparse time-base* of the TTA, explained above, makes it possible to define a system-wide notion of time, which is a prerequisite for an indisputable borderline between the past and the future, and thus the definition of a system-wide distributed state. The "interval of silence" on the sparse time base forms a system wide consistent dividing line between the past and the future and the interval when the state of the distributed system is defined. Such a consistent view of time and state is very important if fault tolerance is implemented by replication, where faults are masked by voting on replicated copies of the state. If there is no global sparse time-base available, one often recourses to a model of an *abstract time* that is based on the order of messages sent and received across the interfaces of a node. If the relationship between the *physical time* and the *abstract time* remains unspecified, then this model is imprecise whenever this relationship is relevant. For example, it may be difficult to determine in such a model the precise state of a system at an instant of physical time at which voting on replicated copies of the state must be performed.

3.3 Simple TTA Nodes

A simple node of the TTA is composed of three subsystems: a communication controller to the time-triggered communication system that contains two independent multicast communication channels, a host computer to perform the application tasks, and a communication controller to access the process input/output (Figure 3).

At the architecture level there are two operationally fully specified (in time and value) interfaces between the three subsystems within a node, called communication network interfaces (CNIs). The CNIs form temporal firewalls that eliminate control error propagation by design [11]. The communication system transports state messages from the CNI in the sending node to the CNIs in the other nodes within a cluster of nodes via the replicated communication channels. Since state messages are not consumed on reading and a new version of a state message overwrites the previous one, the CNI for a state message can be placed in a dual-ported memory. The data flow and control flow between a sending host computer in one node and a receiving host computer in another node is shown in Figure 4. The instants when messages are fetched from the sender's CNI and are delivered at the receiver's CNI are known *a priori* and are common knowledge to all communicating partners within the TTA. These instants establish the temporal specification of the CNIs.

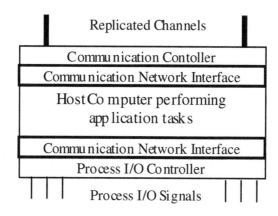

Fig. 3. Internal Structure of a Simple Node of the TTA

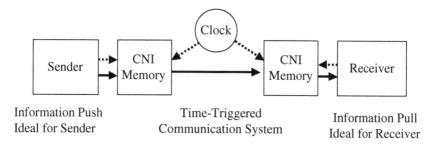

Fig. 4. Data flow (full line) and control flow (dashed line) across a temporal firewall interface

4 Multi-criticality Nodes

The Time-Triggered Architecture is an integrated architecture that provides the framework for the implementation of large embedded systems consisting of diverse distributed subsystems of differing criticality. We have coined the term *Distributed Application Subsystem (DAS)* for such a nearly independent subsystem of a large embedded system[12]. Consider, for example, an automotive control system: there are the power-train control system, the airbag system, the multimedia system of the car, etc., all nearly independent subsystems that interact with each other via a controlled information flow.

We assume that each DAS can be modeled by a set of micro-components (i.e. a hardware/software unit including operating system, middleware, and the application software) that communicate by the exchange of state and event messages across a virtual communication channel of known temporal properties. It must be assured by the TTA that there are no unintended side effects--neither in the domain of time, nor in the domain of values-- between different DASs. In order to meet this requirement we propose a new hardware/software structure for nodes that are implemented on a single giga-scale TTA SoC (system-on-a-chip).

Figure 5 gives an overview of e proposed structure of a TTA SoC. At a high-level of presentation, the SoC can be viewed as containing a set of micro-components that are interlinked by a time-triggered conflict-free network-on-a-chip (NoC). In order to arrive at a *splittable design*, we require that the micro-components operate *nearly autonomously* and interact with each other *only* via well-defined message interfaces.

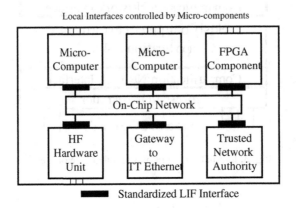

Local Interfaces controlled by Micro-components

Fig. 5. Overview of the SoC

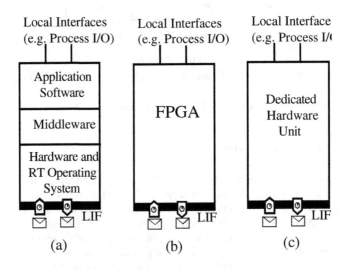

Fig. 6. Differing implementation technologies of a micro-component
(a) programmable computer, (b) FPGA, (c) dedicated hardware unit

The NoC must provide to each DAS a deterministic communication channel with a known bandwidth. A micro-component can be implemented as a programmable computer with its own hardware, operating system and application software (Figure 6 a), as a FPGA module that implements the intended functionality in programmable hardware (Figure 6 b) or a dedicated custom made hardware unit (Figure 6 c). However, all the micro-components of Figure 6 must contain to same interface structure towards the NoC to be TTA conformant. In this new structure the need for a node global operating system that encapsulates the subsystems of each DAS disappears.

The SoC can be characterized by the following properties[13]:

(i) Strict separation of computation and communication: We follow a computational model that partitions the behavior of distributed applications into phases of computation and communication. Computation is performed within the micro-components and communication among the micro-components is realized by a *deterministic time-triggered* on-chip network. The computation phase encompasses the acquisition of information, the processing of information, depositing and retrieving the information in long-time storage, and outputting the information to the environment of the SoC. The communication phase covers the transmission and reception of messages among the micro-components that populate the SoC.

(ii) *Abstraction from the internal structure and behavior of a micro-component:* We introduce the concept of a *well-specified linking interface (LIF)* [14] that describes the behavior (both in the *value domain* and the *temporal domain*) of a micro-component that is relevant for the user and makes no assumption about the implementation technology (hardware/software) of a micro-component, provided the temporal constraints are met. *We standardize the LIF, not the micro-component behind the interface.* A micro-component can be stand-alone computer, an FPGA block or a custom hardware unit that operates at its own *adjustable* frequency that can be different from the frequency of the other micro-components (Fig. 6). From the point-of-view of the user of a micro-component it is sufficient to understand the LIF specification including its interface model; knowledge about the internal structure or behavior of a micro-component is neither required *nor recommended,* since the implementation technology which is hidden behind the *stable LIF* may change as a consequence of technological developments. The precise and stable LIF specification supports the *reuse* of micro-components on different SoCs and within differing application domains.

(iii) *Introduction of a deterministic time-triggered on-chip network for the interaction among micro-components:* The on-chip network provides *conflict-free* communication channels of *a priori known* temporal properties between the LIFs of the micro-components. The on-chip network supports two types of communication: the periodic transmission of state messages and the guaranteed transmission of event messages. The conflict-free bandwidth allocation can be adapted dynamically to evolving demands of the particular application. This re-allocation is performed by a dynamic *trusted network authority* (TNA) that is hosted in one of the micro-components.

(iv) *Continuous monitoring and control of the power-consumption of the SoC and the timeliness of the micro-components by the TNA:* The trusted network authority (TNA) monitors continuously the power of every micro-component and the global power-level of the SoC and integrates the dynamic bandwidth allocation, and the scheduling with dynamic power management of every individual micro-component in order to save energy. In case a micro-component develops a permanent fault, the TNA may be able initiate a dynamic reconfiguration and reallocate the software of the faulty micro-component to a healthy unit.

(v) *Openness to the Internet*: We assume that a generic embedded system architecture must support a secure connection to the Internet. The proposed platform architecture provides potentially two alternatives for an Internet connection: a *wire-*

bound connection via TT-Ethernet gateway micro-component and a *wire-less connection* via an on-chip sender/transmitter micro-component supporting a standard wire-less protocol. In addition to being fully compatible to standard Ethernet, TT-Ethernet supports the deterministic transmission of time-triggered messages. This determinism is needed if we intend to build fault-tolerant systems that mask complete chip failures by triple-modular redundancy (TMR).

5 Fault Tolerance

In a safety-critical application an SoC must be considered to form a single fault-containment region (FCR) that can fail in an arbitrary failure mode. A restricted failure-mode model requires two *independent* FCRs (one FCR monitoring the behavior of the other FCR) which cannot be housed on the same die because of the many common mode elements of a single die, such as: power supply, mask, production process, physical space. We therefore need a deterministic off-chip communication channel, such as TT Ethernet with a special guardian in the switch [15] to provide fault-isolation and error detection in the temporal domain at the architecture level. Our platform SoC architecture, which is based on the TTA [1], performs error detection in the time-domain at the level of the SoC-external architecture and error detection in the value domain by triple-modular redundancy (TMR) .

There is an additional benefit in such an architecture approach if the nodes are formed by giga-scale SoCs. It is expected that in technologies beyond 90nm feature size, single-event upsets (SEU) will severely impact field-level product reliability, not only for embedded memory, but for logic and latches as well [16, 17]. This effect can be mitigated by providing a triple-modular redundant structure, consisting of three SoCs, for masking transient, intermittent, and permanent SoC faults.

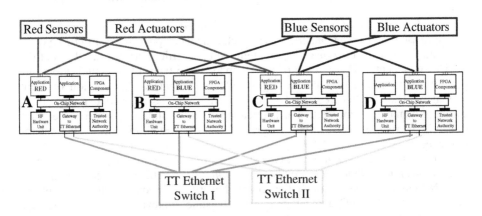

Fig. 7. TMR configuration in the TTA

Figure 7 depicts a triple-modular redundant (TMR) configuration of four SoCs of the platform. Each SoC supports a number of different DASes. Let us assume that the services of two DASes, the red DAS (right) and the blue DAS (left), are safety-critical. We will instantiate the replicated micro-components of these safety-critical DASes on three SoCs (on SoC A, B, and C for the red DAS, and on SoC B, C, and D for the blue DAS) under the assumption that each SoC forms an independent fault-containment region [18]. The communication between the SoCs is realized by two replicated external deterministic communication channels via the TT service of TT Ethernet [15]. It is assumed that the *internal state* of each red DAS micro-component is periodically distributed to the other two red DAS micro-components for the pur-pose of outvoting a transient error in the internal state. The same must hold true for the blue DAS. The duration of the period of the internal state distribution determines the repair time after the occurrence of a transient fault and is a critical parameter of any reliability model. Replicated sensors input the information from the environment to the respective micro-components. The output is delivered to fault-tolerant voting actuators. In the configuration, the failure of any single device (input, output, SoC, and any one of the two communication subsystems) is tolerated. A prerequisite for such a fault tolerant structure to mask an error in any one fault-containment region is the availability of a global notion of time and the timely and deterministic behavior of the communication service among the SoCs and within the SoCs and the processing within the micro-components.

6 Conclusion

The Time-Triggered Architecture provides a framework for the implementation of triple modular redundancy (TMR) such that a continuous service can be provided to its users, even in the presence of arbitrary component failures. In this paper we have elaborated on the fault hypothesis of the TTA, described the structure of the TTA and speculated about the future implementation of the TTA on a single giga-scale system on a chip (SoC).

Acknowledgments

This work has been supported in part the European Integrated Project DECOS under project number IST-511764, by the Austrian FIT IT project on TT Ethernet under project number 808197/1729-KA/HN, by the European Network of Excellence ARTIST II under project number IST-004527. Many discussions with within our research group at the TU Vienna are warmly acknowledged.

References

1. Kopetz, H. and G. Bauer, *The Time-Triggered Architecture*. Proceedings of the IEEE, 2003. **91**(January 2003): p. 112-126.
2. Pauli, B., A. Meyna, and P. Heitmann, *Reliability of Electronic Components and Control Units in Motor Vehicle Applications*. 1998, Verein Deutscher Ingenieure (VDI). p. 1009-1024.
3. Powell, D. *Failure Mode Assumptions and Assumption Coverage*. in *Proc. 22nd Int. Symp. on Fault-Tolerant Computing (FTCS-22)*. 1992. Boston, MA, USA: IEEE Computer Soci-ety Press.

4. Ademaj, A., et al. *Dependability Evaluation of the Time-Triggered Architecture with Bus and Star Topology*. in *DSN Conference*. 2003. San Francisco: IEEE Press.
5. CAN, *Controller Area Network CAN, an In-Vehicle Serial Communication Protocol*, in *SAE Handbook 1992*. 1990, SAE Press. p. 20.341-20.355.
6. Driscoll, K. and K. Hoyme, *SafeBus for avionics*. IEEE Aerospace and Electronics Systems Magazine, 1993. **8**(3): p. 34-39.
7. Thurner, T. and Heiner. *Time-Triggered Architecture for Safety-Related Distributed Real-Time Systems in Transportation Systems*. in *FTCS 28*. 1998: IEEE Press.
8. Hazucha, P. and C. Svensson, *Impact of CMOS technology scaling on the atmospheric neutron soft error rate*. IEEE Transactions on Nuclear Science, 2000. **47**(6): p. 2586-2594.
9. Kopetz, H. *Sparse Time versus Dense Time in Distributed Real-Time Systems*. in *Proc. 14th Int. Conf. on Distributed Computing Systems*. 1992. Yokohama, Japan: IEEE Press.
10. Mesarovic, M.D. and Y. Takahara, *Abstract Systems Theory*. Lecture Note in Control and Information Science. Vol. 116. 1989: Springer Verlag.
11. Kopetz, H. and N. Suri. *Compositional Design of Real-Time System: A Conceptual Basis for the Specification of Linking Interfaces*. in *ISORC 2003--The 6th International Symposium on Object Oriented Real-Time Computing*. 2003. Hakodate, Japan: IEEE Press.
12. Kopetz, H., et al., *From a Federated to an Integrated Architecture for Dependable Embedded Systems*. 2004, Research Repport TU Vienna.
13. Kopetz, H., Ademaj, A., El-Salloum, C., Grillinger,P., and B. Huber, Peti, P., Obermaisser, R., Steinhammer, K.,, *A Time-Triggered SoC Platform for Distributed Embedded Applications*. 2005, Technical University of Vienna: Vienna. p. 17.
14. Jones, C., et al., *DSOS Conceptual Model*. 2003: University of Newcastle upon Tyne, Techn. Report CS-TR-782, TU Vienna, Technical Report 54/2002, Qinetic Technical Report TR030434, LAAS Technical Report. p. 1-122.
15. Kopetz, H., et al. *The Design of TT Ethernet*. in *ISORC 2005*. 2005. Seattle: IEEE Press.
16. Constantinescu, C. *Impact of Deep Submicron Technology on Dependability of VLSI Circuits*. in *Proc. of the 2002 International Conference on Dependable Systems and Networks*. 2002. Washington D.C.: IEEE Press.
17. Roadmap, I., *International Technology Roadmap vor Semiconductors, 2003 Edition*. 2003, Semiconductor Industry Association.
18. Kopetz, H. *Fault Containment and Error Detection in the Time-Triggered Architecture*. in *ISADS 2003*. 2003. Pisa: IEEE Press.

The Value of Conformance Testing and a Look at the SAF Test Project

Bob Spencer

Intel Corporation, 2111 NE 25th Ave., Hillsboro, OR 97124
bob.spencer@intel.com

Abstract. Industry acceptance of the Service Availability™(SA) Forum interface specifications is apparent with the increasing number of commercial and open-source implementations based on the Hardware Platform Interface Specification and Application Interface Specification. To measure completeness and establish a standard, SA Forum is now preparing the process to certify implementations of the B.01.01 specifications and beyond. SAF Test is an open source project where the tests used for certification are created. This paper discusses the value of conformance testing and certification for service availability, especially with regard to the Service Availability Interfaces. It also describes the challenges faced in starting and maintaining an industry-wide conformance test project, and the value SAF Test brings to SA Forum implementations and customers.

1 Introduction

The Service Availability™Forum[1] recently released a new version (B.01.01) of both the Hardware Platform Interface (HPI) Specification and Application Interface Specification (AIS). The new versions have been quickly adopted by open source projects. OpenHPI[2] released a new implementation at the end of 2004, and OpenAIS[3] will complete implementations of many of the AIS services in the first half of 2005. A look at the SA Forum product registry shows that many commercial implementations of prior specifications have already been completed. It is anticipated that many of these will be updated and that new implementations based on the latest version will be completed this year.

Accompanying the success of SA Forum in the industry is the responsibility to certify each of the implementations and provide a measure of correctness to customers. While each implementation may have its own validation tests, this does not ensure that any two interpretations of the specification will be consistent. SA Forum certification, in conjunction with SAF Test, offers a solution. Beginning in the 2nd half of 2005, SA Forum will begin the formal process of certifying HPI implementations based on the B.01.01 specification. AIS certification is anticipated to begin in the first half of 2006. The tests that will be used

---

[1] Service Availability™Forum (http://www.saforum.org).

[2] The OpenHPI Project (http://openhpi.sourceforge.net).

[3] The OpenAIS project (http://developer.osdl.org/dev/openais).

M. Malek, E. Nette, and N. Suri (Eds.): ISAS 2005, LNCS 3694, pp. 15–24, 2005.

for certification are being created and maintained in SAF Test, an independent, open source project. This project provides an open test repository containing all SA Forum conformance tests for any vendor to use, inspect, and contribute to while preparing their implementations for SA Forum certification.

Strict compliance to a specification does not guarantee a fool-proof solution; however it does have many advantages. This paper describes the ways that conformance testing can help a project and where it falls short and also talks about the benefits of certification. In addition, the SAF Test project will be presented with its strengths and shortcomings as a viable solution to providing conformance test suites for open specifications.

2   Conformance Testing

The objective of conformance testing is to establish whether an implementation being tested conforms to the specification as defined in a standard[1]. Conformance testing is generally much clearer than functional or behavioral testing and only limited by the clarity of the specification, or lack thereof, in identifying acceptable parameter values, call order, and return codes. Guidelines, such as those found in the IEEE Standard for measuring conformance to the POSIX standards[2], offer a clear format for conformance tests and identify levels of testing, test assertion documentation, and output format.

2.1   What It Provides

Testing with a complete conformance test suite provides many benefits. First, it is a good exam of the overall software and determines if there are any places the implementation does not follow the standard. It will exercise all methods with a thorough number of possible parameters, including multiple valid values, boundary conditions, and multiple error conditions. Often tests written in-house will do a good job at covering the most-used execution path or easily produced errors but leave many of the less-often-used or less-understood methods under-tested. Second, the rigor of compliance testing is a good stress test and identifies implementations that can be relied upon. Since every method is exercised multiple times, some confidence can be obtained that it won't crash given valid and invalid parameters. Finally, one of the primary reasons for basing an implementation upon a standard is to ensure modularity. Passing conformance testing provides a measure of confidence that the implementation can be used in modular service availability environments. While this last point is not guaranteed, it is unlikely to be satisfied without at least passing conformance tests.

2.2   What It Does Not Provide

Conformance testing alone is not sufficient for commercial-quality solutions. It does not provide any performance measurements. Unless calling a method hangs indefinitely it may still pass conformance testing and be slow or even unusable. It also doesn't test for side effects unrelated to the task at hand, such as deleting the wrong file or consuming too many resources and interfering with other processes. It does nothing for interoperability testing unless specifically called out in the specification. For example, SA Forum HPI implementations may leverage IPMI to gather information from the hardware. During conformance testing the HPI implementation may substitute calls to IPMI with calls to an IPMI stub or dummy plugin. Though the HPI implementation will pass conformance testing there is no guarantee that it will work when using IPMI. Finally, conformance testing does not cover the issues of hardware compatibility which is especially important for hardware specifications such as the HPI specification.

Table 1. Summary of Conformance Testing Benefits

What it can do	What it can NOT do
Shows if the implementation follows the specification	No performance measurements
Provides a good stress of the implementation	No test for undesired side effects
Exercises all the methods with a reasonably thorough combination of parameters	No interoperability check
Demonstrates a level of stability	No hardware compatibility check
Provides a strong assurance of modularity	

Given the unambiguous nature of conformance testing, the results are usually concrete. This concrete result is what certification relies on and certification is discrete: either the implementation is certified, or it is not.

3   Certification

The quality of certification relies on the accuracy and thoroughness of the conformance tests, which as previously mentioned, relies on the clarity of the specification. In the case of Service Availability Interfaces, the clarity and quality has improved through multiple revisions and contributions of leading industry players. This enables a solid suite of tests to be written that will give credibility and meaning to the SA Forum certification. However, wide-spread use of implementations based on these specifications will be the real qualifier.

3.1   Why Get Certified?

"My implementation has been working in-house and at our beta customers' sites for over a month. Why do I need to get certified?"

Given the variety of implementations, it is important to have a formal process which can verify that a given solution fully conforms to the specification. Though well-intentioned developers may do their best to fully comply with the specification, and well-intentioned test writers may do their best to fully test the implementation, errors will still exist due to differing specification interpretations, software bugs, or incomplete testing. Certification is usually a much more thorough examination of the implementation, especially with regard to the less-often-used methods or more obscure parameter value possibilities.

By completing the certification, implementers benefit in multiple ways. First, they have one good measure of the completeness of their code. Second, they can claim their solution is certified by SA Forum to customers and competitors. Finally, they benefit from the increased value of the specification. Without certification it is impossible to ensure conformance and without conformance the specification becomes a suggestion instead of a standard. Basing an implementation on a standard interface increases the likelihood that it will get used.

4   SAF Test

The SAF Test open source project was first introduced in September, 2004 with a mission to become the central repository of conformance test suites for SA Forum published specifications including HPI and AIS. Given that published SA Forum specifications are open, it made sense for the conformance tests to be open too. Another influence in starting the project was the successful track record of the Open POSIX Test Suite which contains conformance tests for the IEEE Std 1003.1-2001 System Interfaces specification and has been running since 2002.

SAF Test is setup on sourceforge where the mailing lists, CVS-based source repository, and web site are maintained (http://saftest.sourceforge.net).

4.1   Project Challenges

Setting up the SAF Test project went fairly smoothly as it had the support of key SA Forum member companies. Here are some of the challenges we faced, and resolutions if they've been resolved.

– **Licensing:** One of the first big challenges was deciding what open source license to use. Some companies desired a BSD license so that proprietary additions (such as functional tests) could be distributed with the SAF Test suite to customers without having to give those changes back to SAF Test or the general public. Other companies felt that conformance tests, and all changes and enhancements, should be open for everyone to use. Allowing companies to take the conformance tests and modify them for their own needs would potentially create multiple versions of the tests and reduce the value of a single open repository. In the end, the GPL license was selected. We are hopeful that this decision will not prevent some partners from using the tests.

- **Test Collection:** When we began, many implementations of the HPI and AIS specifications already existed, both proprietary and open source. The first objective of the group was to collect all the valuable test cases and where possible have the owning companies release them and port them to the new framework. There were hundreds of tests at various stages of completeness. Although this work was somewhat tedious, in the end the ability to import these tests gave the project a good start and also helped involve others more quickly.
- **Licensing Tests to Import:** Including tests from other projects always involves checking for license compatibility. In a couple of cases we had to wait for legal departments to approve the release of their companies' tests. For example, OpenHPI had already completed the first implementation of the SA Forum HPI A.01.01 specification when SAF Test was started. There were hundreds of excellent conformance tests that were part of the project that SAF Test wanted but needed re-licensing. Eventually the tests were opened for GPL license and imported.
- **Open contributions:** Open source often means open for criticism. While we have not had great challenges in this regard, it is more difficult to manage change in the open source than in an in-house project. Final say usually takes community agreement. However, this is also the advantage of having solutions open.
- **Limited contributions:** Community involvement in the development work is isolated to a small number of people from a couple of companies. While this is understandable, it is still challenging to get the work done with just a few resources.

4.2 Benefits of SAF Test

The value of SAF Test has been seen in many areas and offers benefits to both SA Forum as well as specification implementers:

- **A single test repository:** All conformance tests for SA Forum specification are organized, always available, and in a well-known location. This, in turn, has benefits, namely:
  - **Eliminates duplicate test creation:** Each company that decides to implement one of the Service Availability Interfaces knows that they already have a complete conformance test suite at their disposal. This reduces duplication of work and saves them money.
  - **Consistency across the industry:** Every company has the same opportunity to test their implementations against the standard.
- **Test creation is independent of any particular implementation:** With an open door for solutions, suggestions, and contributions, the tests are not just tailored for a particular solution. In addition, the beneficiaries are not just the open source implementations. Whether or not companies are contributing resources, they are still able to utilize the tests.

– **Tests and framework can be reviewed and corrected by anyone:**
In relation to being independent, the framework can be seen by anyone and
updated publicly.
– **Tests and framework become hardened:** With multiple vendors using
the tests and verifying their solutions, the tests become hardened before they
are used in the certification process.
– **Certification is more transparent:** Implementations can be pre-verified
by creators, saving time and money. There should be no big surprises when
it comes time for official certification.
– **SA Forum saves money while maintaining control over the out-
come:** Since contributions come freely from participating companies, SA
Forum does not need to invest as much in the development of the tests. At
the same time, SA Forum working groups can monitor the progress and can
insert ideas and make clarifications as the work progresses. Finally, the tests
will be reviewed by SA Forum before certification begins.
– **Communication:** SAF Test developers, the SA Forum working group
members, and the community have a forum to discuss issues relating to
the specification and conformance. There have already been many questions
posed regarding what a particular portion of the specification means or why
a test returns the values it does. If necessary, SA Forum members can take
specification bugs back to the working group for changes.

4.3   Test Framework

The test framework contains many of the conformance test standards specified
in the IEEE Standard for measure conformance to the POSIX standards. As it is
possible to get a more up-to-date description of the current framework by reading the online documentation on the SAF Test web site, only a brief overview
will be provided here.

The test suite can be downloaded from the SAF Test web site. It contains
a simple directory structure that contains the tests and execution scripts. The
layout is shown in Figure 1.

Running make will build all tests in the current directory and sub-directories.
The script run_tests.sh executes all tests in the current directory and sub-directories. The output is a file called result.txt that contains the number of tests that
passed, failed, blocked, or were not supported.

In the interface folders (see saCkptInitialize in Figure 1), tests are identified
by 1.c, 2.c, ... The file assertion.xml describes each of the tests and their expected
return code. This information is also contained within the test file header. The
file coverage.txt identifies which tests are still needed for complete conformance
coverage.

Conformance. All tests in the SAF Test suite are simple conformance tests. As
described above, conformance tests verify that every method in the
specification is implemented, can be called successfully, and returns the correct value given correct and incorrect parameters. Each specification directory

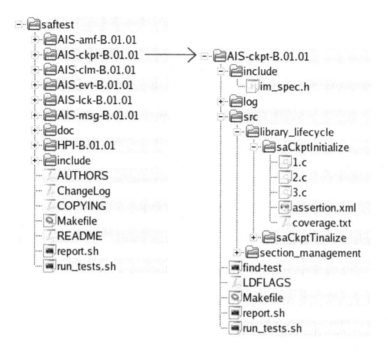

Fig. 1. Layout of SAF Test directories

(see AIS-ckpt-B.01.01 in Figure 1) will contain a description of all test cases, along with tests that validate output of each method given all possible (within reason) combinations of input and output. The test suite will provide line-by-line coverage of the SA Forum B.01.01 specifications and any future specifications used for certification. This coverage will follow the guidelines for Thorough Testing as stated in section 7.2.2 of the IEEE Standard for measuring conformance to the POSIX standards:

> Thorough testing is a useful alternative to exhaustive testing. Thorough testing seeks to verify the behavior of every aspect of an element but does not include all permutations. For example, to perform thorough testing of a given command, the command shall be tested with no options and then with each option individually. Possible combinations of options also may be tested. [2]

Each API will be exercised multiple times given parameters that cover normal and error conditions, including boundary conditions if it applies. In addition, methods that accept multiple parameters will be exercised with a reasonable number of value combinations (correct and incorrect) to verify that the predicted results are generated. If the order in which API's are called causes different results, an attempt will also be made to exercise these possibilities.

Although this level of testing is not absolutely comprehensive, it can still provide sufficient confidence that a given implementation is complete and will work.

Licensing. As mentioned previously, all tests and documentation in SAF Test are licensed as GPL, version 2. The GPL license was selected because of the nature of the tests, with the basic position that conformance tests should be open and that changes and enhancements to the test suite should be made available to everyone.

4.4   Brief Project Status

The first goal of SAF Test was to complete all tests for the HPI and AIS A.01.01 specifications by the end of 2004. The project enjoyed a quick start by leveraging tests that had already been developed for OpenHPI and LinuxHA. The tests were organized into a single, well-structured framework. This framework and related documentation simplified test contribution and made it easy for developers to see what areas were not yet complete. Scripts were created to automate the building and execution of the tests.

In mid-January all HPI tests, and all AIS tests except AMF were completed for A.01.01. These tests have been used by OpenHPI, OpenAIS, LinuxHA, and others to validate their implementations as well as prove the test framework and identify areas for enhancement.

Note: SA Forum will not be using the A.01.01 tests or certifying A.01.01 implementations. However, completion of the A.01.01 tests satisfied the intended goal of solidifying the test framework and motivating implementers to contribute to the project and participate on the mailing lists.

SAF Test will provide the tests for SA Forum certification which begins with the B.01.01 specification. Beginning again with an excellent base of tests from the OpenHPI project (~500 tests), the test suite for HPI should be completed by the end of Q2, 2005 (>700 individual tests). SA Forum is planning to begin certification for HPI implementations in early Q3, 2005. AIS test creation is also ongoing, and tests for individual services will be made available as they are completed. All AIS tests (>1500 individual tests) should be completed before the end of 2005.

SAF Test will continue to provide tests for future SA Forum specifications and versions. In addition, more in-depth, or functional testing may be added to enhance the conformance tests, as requested by SA Forum.

4.5   SAF Test Schedule

The schedule for SAF Test is influenced by the number of resources contributed by participating companies. The primary milestones are listed in Table 2. For more details, please visit the project website.

Table 2. SAF Test Schedule

Date	Task	Status
Jan. 15, 2005	**AIS-A.01.01 and HPI-A.01.01 Complete** **Definition:** Release AIS A.01.01 services except AMF. Release HPI A.01.01 tests.	Complete
Feb. 11, 2005	**AIS-B.01.01 Membership** **HPI B.01.01 test coverage documented** **Definition:** Release AIS B.01.01 membership service line-by-line tests. Document how to get line-by-line coverage of AIS and HPI B.01.01 specifications. Import OpenHPI B-spec tests into SAF Test framework.	Complete
Apr. 06, 2005	**AIS-B.01.01 Checkpoint Service** **HPI-B.01.01 General and Domain Sections** **Definition:** Release AIS B.01.01 checkpointing service tests. Release HPI B.01.01 line-by-line tests for the general and domain sections.	Complete
May 15, 2005	**HPI-B.01.01 Sensor, Control, Inventory, Hotswap** **Definition:** Release major portions of the HPI B.01.01 resource section including sensor, control, inventory and hotswap tests.	Complete
May 20, 2005	**AIS-B.01.01 Event Service** **Definition:** Release AIS B.01.01 event service line-by-line tests.	Complete
May, 2005	**SA Forum Working groups begin review of HPI-B.01.01 tests**	Not started
June 15, 2005	**HPI-B.01.01 Complete** **Definition:** Complete line-by-line coverage for complete HPI B.01.01 specification. Tests are ready to hand over for SA Forum certification.	In progress
July 2005	**SA Forum Working groups complete review of HPI-B.01.01 tests**	Not started
July 2005	**SA Forum Certification begins for HPI B.01.01**	Preparation started
December 2005	**AIS B.01.01 Complete** **Definition:** Complete line-by-line coverage for complete AIS B.01.01 specification.	Not started

5 Summary

Conformance testing hardens an implementation of a standard and provides a good measure of confidence about the implementation's ability to be used in a modular, service availability environment. Additionally, the tests provide a degree of stress testing and thorough coverage of the code. Once an implementation is able to successfully complete a full conformance test pass, it is advantageous to the implementer and the customer for the implementation to become certified.

In the case of SA Forum certification, the conformance test suite that will be used will be the one developed in the SAF Test open source project. This open development has numerous benefits and very few disadvantages. If anyone is preparing a solution that implements one of the Service Availability Interfaces, they should become familiar with the SAF Test project and take advantage of the excellent source of conformance tests. SA Forum certification is just around the corner. Climb aboard!

References

1. Tam, F. and Ahvanainen, K., *First Experience of Conformance Testing an Application Interface Specification Implementation*, Service Availability, First International Service Availability Symposium, ISAS 2004, Revised Selected Papers, Munich, Germany, May 13-14, 2004.
2. *IEEE Std 2003-1997: Requirements and Guidelines for Test Method Specifications and Test Method Implementations for Measuring Conformance to POSIX Standards*, 1998.

Building Highly Available Application Using SA Forum Cluster: A Case Study of GGSN Application

Ajay Kamalvanshi[1] and Timo Jokiaho[2]

[1] Nokia Corporation,
313 Fairchild Drive, Mountain View, CA-94043, USA
ajay.kamalvanshi@nokia.com
[2] Linnoitustie 6, 02600 Espoo, Finland
timo.jokiaho@nokia.com

Abstract. Most of the highly available applications are built on top of expensive cluster software with proprietary application interface and have problems with maintenance and enhancements. The SA Forum has standardized application interface for building highly available applications. This paper discusses our experience with early adoption of SA Forum specification both for implementation of a cluster as well as developing a real-world telecom application. This paper outlines steps to develop highly available application on a SA Forum based cluster using an element of 3G networks, GGSN, as a case study. Various design considerations such as building standby; synchronizing between active and standby; and handling switchover are discussed in detail. In addition, we have included the lessons that we learnt during development, integration, and testing to help prospective developers of complex real-time telecom applications.

1 Introduction

The SA Forum has defined Application Interface Specification (AIS) for developing highly available applications particularly for the telecom domain where the downtime requirements are less than six minutes in a year [8]. This paper discusses experience of developing a Highly Available (HA) middleware and a telecom application based on SA Forum's specifications. The cluster is implemented on a multi-blade multi-chassis system to provide third generation (3G) network functionality. The system is part of the network architecture as proposed in Third Generation Partnership Project's (3GPP) release 5 specifications and functions as a Gateway GPRS Support Node (GGSN).

This paper begins with an overview of the target platform and its software architecture. The overview is then followed by introduction of HA middleware that implements the SA Forum cluster. Subsequently, the paper focuses on the application. A brief introduction covers the 3G domain and application's functions. The remaining paper covers various high availability aspects of the GGSN application.

M. Malek, E. Nette, and N. Suri (Eds.): ISAS 2005, LNCS 3694, pp. 25–38, 2005.

2 Overview of Cluster Platform

The cluster platform consists of two chassis containing several computing blades, as illustrated in Fig.1. Logical overview of cluster hardware. Each blade in both chassis is a fully functional unit in terms of hardware and operating system, i.e., each blade has processor(s), memory, and an instance of an operating system. The blades are further connected with each other using switched redundant Ethernet for inter-blade communication. Based on the provided system functionality, the chassis are referred to as Server and Router chassis. The server chassis provides value added services or control plane functionality. The router chassis provides routing and forwarding functionality, also known as user plane functionality. The line card interfaces that connect to other equipments in the 3G networks are located on router blades.

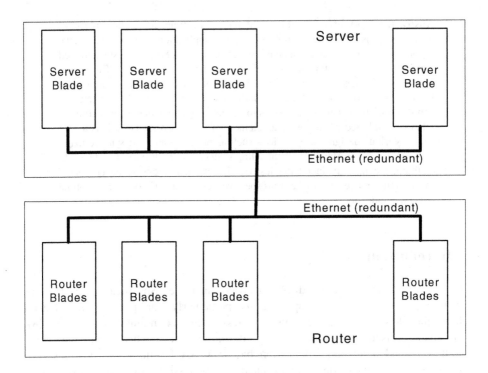

Fig. 1. Logical overview of cluster hardware

The GGSN application described in this paper runs on server blades and uses router blades for external connectivity and setting up user specific forwarding table.

2.1 Software Architecture

The telecom operators demand network equipment to be scalable -- in short term and long term -- as well as highly available. In hardware, this translates to designing extendable system beyond one chassis, and on chassis-level, support for hot-swappable blades. In software, this translates to scalable distributed systems that are generically referred to as clusters in computing parlance.

Nokia has built an Intelligent Network Operating System (INOS) to address distributed carrier-grade requirements for networking equipments. The software architecture, as depicted in Figure 2 Software architecture, is layered to provide flexibility. Each component in the architecture provides services based on well-known interfaces. Thus, any component could be independently replaced by competing components that provide similar interface with controlled impact on overall system. The INOS uses a carrier-grade base operating system for basic services such as scheduling, process management, memory management, I/O, and network services. The shelf management component abstracts hardware management aspect of chassis.

The system management module, distributed across various blades (and chassis), provides unified single system view. The clustering middleware provides infrastructure for service availability via SA Forum defined AIS interfaces. Finally, the applications are built on top of INOS with standard service interface such as POSIX and AIS.

The communication among various nodes of the cluster is based on fast message queue based Inter Process Communication (IPC).

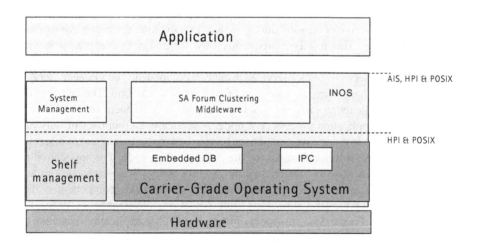

Fig. 2. Software architecture

The rest of this paper is focused on application and service availability middleware.

2.2 Service Availability Middleware

The service availability framework of the INOS is based on the SA Forum's AIS A 1.0 specification [9], although our initial implementation started with draft version (0.8), to achieve fast adoption and possibility to provide real-life feedback to the SA

Forum. We selectively adopted few areas of specification due to product release deadlines and to prevent conflict with our existing stable services. We started with Application Management Framework (AMF) and peer-to-peer check pointing APIs – which were withdrawn from the SA Forum's specification with an understanding to replace it with more generic messaging API in future (1.0 release); however, it was sufficient to address most of our application's need. The AMF is implemented as distributed framework: An AMF process runs on every node and communication among AMF processes is done using IPC. We also decided to continue using our existing message based IPC, name lookup service, and startup procedure, as these were being drafted and reviewed in the forum.

In addition, although designed to be compliant with SA Forum's specification, our implementation is done in phases to support the redundancy model as per product requirement of 2N redundancy – where each active component is protected by a similar standby component.

To realize SA Forum's entities, a mapping is done from existing INOS entities to SA Forum's entities as depicted in Table 1: Mapping of INOS concepts and SA Forum. These are presented here to draw comparison and to provide an example of implementation. As per SA Forum specification, a cluster is collection of cluster nodes. In our system, a cluster node as well as physical node maps to the server blade shown in Figure 1 Logical Overview of Cluster Hardware. Logically, a node represents physical blade along with an instance of operating system, INOS. A node is assigned a role of active or standby during system initialization through hardware and software arbitration. This role is subsequently used for assigning component service instance and HA state to components of a node. The mapping of SA Forum logical entities such as service units, service instance, component, component service instances, and service groups are also simplified for quick adoption. The components are mapped to applications such as GGSN application. A component service instance is assigned to an application process on a node. A service unit comprises of only one component; service instance and component service instance are assigned to a service unit along with HA state based on role of the corresponding node. The application on a node uses the HA state for differentiating the processing. The service group and protection group are formed between service units and components. The corresponding system model is shown in Figure 3 SA Forum System Model entities.

In Figure 3 SA Forum System Model entities, the GGSN application is shown as SA Forum component C1 and C2 on Node 1 and Node 2 respectively. The service units S1 and S2 have only one component C1 and C2 respectively, and form a service group SG1. A service instance A is assigned to service unit S1 and S2 by assigning active HA state to C1 and standby HA state to C2. A protection group, PG A1, is formed is formed between C1 and C2.

When a node with active role fails, the readiness state of its service units is transitioned to out-of-service. This also results in transitioning of associated components' readiness state to out-of-service. The node with standby role then transitions to active triggering transition of its component's HA states to active. After the fail-over the component assumes role of failed component and processes subsequent requests. This is further described in next few sections.

Table 1. Mapping of INOS concepts and SA Forum concepts

INOS concepts	SA Forum concepts
Router and server chassis with multiple blades	Cluster
INOS Node: A computing process complex in a blade with a CPU, memory and I/O along with instance of operating system	Node
Application such as GGSN application	Component
An instance of application on a node such as GGSN process	Component Service Instance assignment
GGSN application along with management interface	Service Unit
GGSN application on a node handling predefined Service workload	Service Instance assignment
Association of GGSN process with active and standby role	Service Group
Association of GGSN process with active and standby role for a workload	Protection Group

Our middleware was required to support two additional features: hot standby and non-revertive switchovers. To support hot standby, all components with standby HA state are always kept in sync with components with active HA state. The primary goal of this scheme is to minimize the affect of switchovers on the user services. Non-revertive switchover refers to the ability of system to prevent unnecessary switchover when the failed active component becomes ready again.

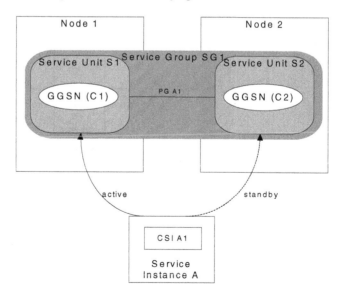

Fig. 3. SA Forum system model entities

INOS also supports in-service upgrades as required by carrier-grade network equipment, but is not discussed here to keep this paper focused.

3   Overview of GGSN

The cluster mentioned in the above sections functions as a Gateway GPRS Support Node (GGSN) in 3G Network Architecture as defined in 3GPP Release 5 specifications: [1], [2], [3], [4], [5], [6], and [7]. Simplified reference architecture for packet switched network is illustrated in Fig. 4. 3G Packet switched network. The figure illustrates position of GGSN in a 3G network. The User Equipment (UE) initially establishes a logical connection or attaches with Serving GPRS Support Node (SGSN) through Base Station (BS) and Radio Network Controller (RNC). The SGSN validates the UE and performs security functions such as authentication and encryption. When UE needs data communication, a Packet Data Protocol (PDP) channel is activated which results in setting up an association between SGSN and GGSN. The association is referred to as PDP context, and upon its successful activation, the setup required for UE to connect to data network is performed. The UE can then seamlessly behave like a network terminal. The details of setup and 3G network architecture are described in 3GPP documents in reference section. The charging gateway shown in the figure collects statistics and converts them into billing information [7]. For the rest of this paper, we will focus on GGSN.

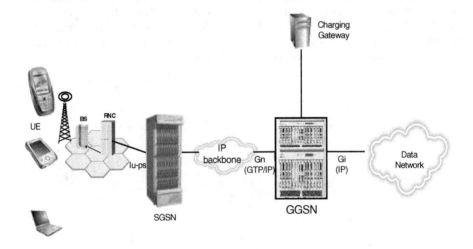

Fig. 4. 3G Packet switched network

The GGSN serves as a gateway between access operator network (also known as Public Land Mobile Network (PLMN) and IP networks such as Internet. All the user packets enter and leave PLMN through GGSN. One or more SGSN are connected to GGSN through Gn interface using GPRS Tunneling Protocol (GTP). GTP is a tunneling protocol for transferring user data with operator network, sometimes known as IP backbone network. GGSN is also responsible for assigning an IP address for the UE that is known to data network. Consequently, it must maintain a routing table for all UEs and route data to the SGSN associated with the UE. The interface between GGSN and data network is referred to as Gi interface and uses Internet Protocol (IP).

The following GGSN functions are supported in our system:

- ❑ Session Management: PDP context creation, context modification, and context deletion.
- ❑ WLAN connection management
- ❑ User profile management
- ❑ Service configuration
- ❑ IP address allocation
- ❑ Subscriber authentication
- ❑ Charging
- ❑ Misc. functions like Operation and Maintenance (O&M) functionality, Fault management, and trace management.

The details of these functionalities are covered in the 3GPP specifications [1], [2], [3], [4], [5], [6], and [7].

3.1 Application Design

The GGSN functions are implemented as an application process with many subsystems. The application process is a Unix daemon that listens on multiple sockets for application events or messages. When an event is received, the registered event handler is invoked. The event handlers are implemented in subsystems and are registered with the main application during startup.

The main subsystems of GGSN application are: session management, charging, statistics, QoS, and access point management. The session management subsystem is responsible for handling GTP message processing for tunnel management such as PDP context creation, deletion, and modification; managing path connectivity with associated SGSN -- sending periodic keep-alive requests, echo requests, towards connected SGSN and responding to keep-alive requests from SGSN (referred to as path management in 3GPP TS29.060 [2]); handling error conditions such as undeliverable user data packets, G-PDU; and handling overload conditions. The charging subsystem implements charging functionality, as defined in 3GPP TS32.215 [7], and processes the GTP' messages. The statistics subsystem is responsible for collecting and maintaining counters. The QoS subsystem is responsible classifying services and mapping services to IP's differentiated service based on traffic classes. The access point management subsystem manages access points – access points identify external network and optionally service to be offered; handles address allocation; and implements authentication.

4 GGSN Application HA

This section explores procedure for making GGSN application highly available. The steps described in this paper may be applicable to other application designed for protocol processing. However, the implementation choices and redundancy models often depends on the product requirements.

The service availability product requirements for GGSN applications are:

❑ The product must support five nines (99.999%) service availability, i.e., planned as well as unplanned downtime should not exceed more than 5.24 minutes in a year.
❑ The product should not have any single point of failure. All the components must be protected by similar components.
❑ Upon failure detection, the switchover time for the user services should not be more than 50 milliseconds (ms).
❑ The remote nodes should not notice failure of the GGSN application during switchover.
❑ All the important statistics and counters such as remaining prepaid time should not be impacted.
❑ The management connectivity such as management telnet session may not be highly available.
❑ Synchronization with standby must have minimum impact on service processing on active.

These requirements map to 2N redundancy model where every service unit is protected by another service unit of same type. We have further extended the protection at node level where every node is protected by similar node by assigning roles based on node's role. This also means that all important states of active and standby are synchronized; and the standby is always ready to assume role of active. The component capability model chosen for GGSN application is 1_active_or_1_standby.

The task of making GGSN application highly available starts with identifying the following aspects of application:

❑ Application startup and role determination.
❑ Association between active and standby components.
❑ Handling switchovers
❑ Reporting errors
❑ Determining synchronization mechanism for building standby including initial warming and updating for dynamic changes.

Application Management Framework (AMF) handles all above aspects but for synchronization automatically. Determining synchronization mechanism requires detailed analysis of application behavior. To help application designers, INOS middleware provides guidelines and classifies schemes under the following categories:

❑ **Static scheme:** This mechanism is useful for applications that don't change behavior after initial configuration. The initial configuration of these applications is done either using run time parameter or a configuration file. This scheme can be easily realized by communicating the configuration information.

❑ **Provisioning based configuration scheme:** This mechanism is useful when operator needs flexibility of changing application states using command line or similar interface. The state or configuration change is reflected on the standby upon completion of the command. Whether the standby immediately updates its states or waits for the role transition is application specific.

❑ **Network or dynamic scheme:** This mechanism is applicable in scenarios when the application is processing network based requests for establishing sessions or populating tables required for forwarding, routing, etc. It enables the state-full replication between active and standby component service instances. In this case, standby is updated after end of every transaction.

In GGSN application, all the above scenarios are used. The initial configuration is synchronized; the operator initiated command line changes are dynamically updated to standby; and GTP processing to create, destroy and modify tunnels is replicated using dynamic scheme. The synchronization mechanisms are illustrated in Figure 5 Synchronization architecture with sequence numbers. The sequence numbers are shown for processing of a typical provisioning command. When a configuration change request is received from the operator, the changes are first made to the database that stores the configuration (step 1 through 3) after validating the request. The database replicates the data on the standby node using peer-to-peer check pointing service (step 4 and 5). The application then changes the dynamic state (step 6). At this point, the application must reflect the same changes to the standby component. Again it uses the peer-to-peer check pointing (step 7 through 11) to synchronize application states. The AMF in the figure is responsible for initial synchronization and communicating availability state changes. For processing network requests such as GTP messages, steps 1 through 5 is skipped.

Observe that database as well as application uses peer-to-peer checkpoint service. The database uses peer-to-peer checkpoint service for replicating persistent information or configuration data, while the application uses it for runtime state synchronization.

4.1   Interaction Between Application and AIS Components

This section describes the interaction between INOS's HA middleware and GGSN application. The timeline is captured in Figure 6 Interaction between GGSN and HA Middleware. The important steps are:

1. The HA middleware determines the node role through hardware arbitration; this role is then assigned to all the components of that node.
2. The GGSN application initializes AMF and peer-to-peer checkpoint areas along the callbacks.
3. The GGSN application then registers itself with the middleware using the handle and the component name.
4. The selection objects are then obtained from the middleware for each area.

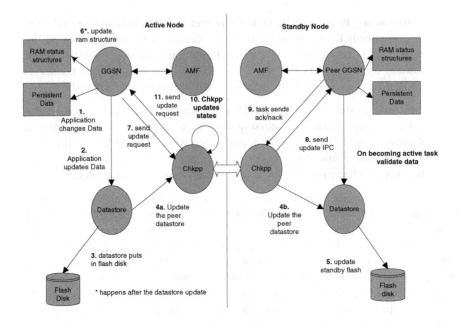

Fig. 5. Synchronization architecture

5. The application requests for passive monitoring and health check start. The handling for failure is currently defined in startup configuration file.

6. The HA middleware (AMF) then calls smAmfCSISetCB to transition the components HA state. In our implementation, this is assigned to role of the Node.

7. The application then registers for tracking changes and responds the periodic health check callbacks. At this time, the application is ready to perform target service. Building of standby is considered to be passive or in background, and the remote node does not know anything about standby. The active component then waits for standby for synchronization steps. All the steps (1 through 7) occur on all the nodes.

8. When standby component is ready to synchronize, it requests for bulk update, or warming request, to active component using saChkppWarmSyncStart. This message is sent to AMF on active that in turn invokes saChkppWarmSyncStartCB callback method.

9. The active component then uses saChkppPush to send data to standby. In the last packet, the active sets no more data flag. When standby receives this, it calls saChkppWarmSyncEnd. At this time, AMF knows that both the components have synchronized state.

10. After this step, the active updates standby whenever important state change occurs on active component.

11. When AMF detects failure of active node, the AMF on standby notifies to all the components to transition from standby to active HA state.

Note that the standby initiates saChkppWarmSyncStart unlike the current checkpoint service where active initiates checkpoint creation. This design decision was made to reduce burden on active and the standby was responsible for initiating synchronization and declaring itself hot-standby.

4.2 Design Description

As mentioned in earlier section, the synchronization mechanism uses the peer-to-peer check pointing APIs that were retracted later. These APIs are very similar to the current messaging APIs. Since the check pointing APIs were being defined when our implementation had started, we decided to use peer-to-peer check pointing. Some of these issues discussed in this section are already addressed in the official check pointing APIs, the new applications are recommended to use check-pointing APIs for future usage.

Handling of Timers. Most of the timers are started only on active to reduce the overhead and synchronize state change between active and standby. If the active and standby change states based on timers independently, then active and standby will soon have different states. However, some exceptions are made for timers that are required for ensuring correct behavior of application after switchover. The examples of these are session timers, DHCP lease timers of IP addresses, and transaction timers.

4.3 Warm Update

Warm update, also called bulk update, is initiated when standby component is ready to synchronize with active component. This transfer is done in background to minimize impact on service processing on active and requires sending the all the relevant states to the standby including:

- ❑ Access Point Database containing all access points
- ❑ IP Addresses
- ❑ Session and lease timers of IP address
- ❑ Configuration data
- ❑ Charging related information such as last record sent out and pending records.

The warm update is always done in a sequence to preserve dependencies across subsystems. Additionally during warm updates, the active component continues to process external events, and all such events are handled as transactions described in previous section.

4.4 Switchover

The HA middleware handles two types of switchover scenarios. The first scenario, graceful shutdown, is initiated upon receiving operator's command. In this scenario, HA middleware sends a notification to all the components in a node. An attempt is made to flush the transient sessions. The GGSN application receives this notification and performs housekeeping function such as informing other components in the cluster; it then completes the processing of all pending configuration transaction. Finally, the standby component is notified, and it validates, flushes incomplete transaction, and assumes active role.

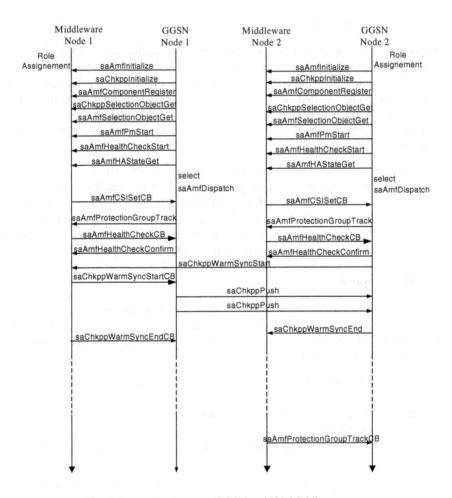

Fig. 6. Interaction between GGSN and HA Middleware

The other type of switchover, forced switchover or sudden death, occurs when one or more components or when a node fails. The failure of active node is detected when standby receives an interrupt indicating failure of active or when it stops receiving heartbeat from active. In this scenario, the standby component is informed about the transition. The standby component does minimum validation before assuming active role.

In both scenarios, external interface for management such as management telnet session, charging gateway connectivity is lost. The connection with management nodes and charging gateway are re-established after the switchover. The data-path traffic, however, continued as the data plane is also synchronized separately with standby data-plane blade. This mechanism ensures the switchover time to be less than 50 ms for data-plane.

5 Lessons Learned

Early adoption always has unforeseen challenges, and simplifying assumptions helps to meet the product milestone. Our decision for simple mapping of node, service unit, service instance, component etc helped us to shorten the development time and testing complexity. Furthermore, our decision to use AMF and peer-to-peer check pointing proved to be sufficient to implement all service availability feature requirements for the product. As in all practical projects, valuable lessons are learned during development, integration, and system testing. Here are some of these lessons that made us wiser:

1. There is nothing like same-size-fits-all redundancy solution. Each application must be studied for external dependencies, the synchronization schemes, and switchover behavior. Use of schemes defined in section 4 were very helpful for new HA developers.
2. Choose synchronization judiciously. While analyzing performance during product integration, we found that many synchronizations messages were unnecessary. Not all the internal states are required to be synchronized with standby.
3. Aggregate small synchronization messages into large message whenever possible.
4. Design for timers carefully. Most of the timers need not be sent to standby; only the timers that are required for external dependencies must be synchronized.
5. All components must validate upon switchover to prevent system instability.

6 Conclusions

There are several HA middleware that provide proprietary solution for building highly available applications. Many of these are often incomplete and cannot be adopted for building real-world telecom applications without substantial modifications. While developing on proprietary solution can be frustrating experience for engineers, the maintenance and enhancements is also expensive. In addition, these proprietary HA solutions cannot be ported to different hardware and software platform easily. SA Forum's AIS has standardized the cluster interface for developing highly available application.

In this paper, we discussed our experiences with using SA Forum's AIS specification both for implementing a cluster as well as developing a real-world application. We have also presented some basic assumptions that helped us to adopt the specification without missing product milestones.

Our experience in making GGSN application highly available using SA Forum specification has been challenging, but successful. The GGSN application is challenging because of hot-standby requirement to prevent loss of existing sessions. With about 20-30% overhead in performance, we saw flawless switchover when active node failed, often less than 30 ms of traffic was lost. The outline of making application redundant will serve as a guideline for prospective developers in considering design trade-offs and reducing time to develop telecom application that demand 99.999% availability.

References

1. 3GPP TS 23.060 General Packet Radio Service (GPRS) Service description; Stage 2, v5.4.0.

2. 3GPP TS 29.060 GPRS Tunneling Protocol (GTP) across the Gn and Gp Interface, v5.4.0

3. 3GPP TS 29.061 Interworking between the Public Land Mobile Network (PLMN) supporting Packet Based Services and Packet Data Networks, v5.4.0

4. 3GPP TS 32.215 Charging data description for the Packet Switched (PS) domain, v5.2.0

5. 3GPP TS 23.107 Quality of Service (QoS) concept and architecture, v5.7.0.

6. 3GPP TS 29.208 End to end Quality of Service (QoS) signalling flows, v5.3.0.

7. 3GPP TS 32.215 Charging data description for the Packet Switched (PS) domain, v5.2.0.

8. T.Jokiaho, F.Herrmann, D.Penkler, L.Moser: "Application Interface Specification of the Service Availability Forum", pp 14-16, Boards and Solutions Magazine, June 2003.

9. SA Forum Application Interface Specification AIS B.01.01

Using Logical Data Protection and Recovery to Improve Data Availability

Wei Hu

Oracle Corporation, 400 Oracle Parkway, Redwood Shores, CA 94065, USA
wei.hu@oracle.com

Abstract. Data availability is crucial to overall application availability. This paper describes data failures and classifies them as either physical or logical. Physical data failures such as data loss and corruptions are introduced in the I/O layers; logical data failures are introduced in the application layer. Current data protection techniques are geared towards physical data failures. This paper reviews physical data protection techniques and their limitations. It then introduces the concept of logical data protection and shows how applications such as the Oracle database can use application knowledge to implement logical data protection and recovery that are more effective than conventional physical data availability technologies.

1 Data Availability

An important aspect of high availability is data availability. *Data availability* is the extent to which an application's data is available and correct – i.e., in a form that allows the application to function. Data is not available when it is destroyed, inaccessible, lost, physically corrupted, or logically corrupted. An application cannot function without its data; highly available systems must therefore address *data failures*. Data failures (or unavailability) can be caused by hardware failure, software failure, human error (which includes malicious attack), and site failure.

Hardware failure can cause data loss and data corruption. Disk failures can cause data loss. Data corruptions can occur when the hardware incorrectly modifies data that is written or read back. Data corruption can be caused by faulty hardware and firmware. Uncorrected memory errors can corrupt the I/O buffers. Even faulty cables can corrupt the data that is transferred. In addition to *improper modification of data*, other variations of corruptions are: *misdirected write* where a data block is written to the wrong location and *lost write* whereby the storage subsystem acknowledged the completion of a write that was actually not done.

Software failure can also cause data loss and data corruption. Every data failure that is caused by hardware can also be caused by software. For example, bugs in file system and I/O code can lose writes, overwrite parts of the data that is being written, or write data to the wrong location. Another example of a software corruption is the *stray write* in which data structures in memory are overwritten by another thread. As the amount and complexity of software in the I/O subsystem grows, the probability of software-induced data failures also increases.

M. Malek, E. Nette, and N. Suri (Eds.): ISAS 2005, LNCS 3694, pp. 39–51, 2005.

Human error can account for up to 40% of unplanned application downtime [1]. Human error can cause all the data loss and corruptions caused by hardware or software. For example, a configuration error such as setting up a swap file over the same disk partitions that hold application data can cause data to be overwritten. This has the same effect as a corruption caused by misdirected writes.

A *site failure* is when an entire data center is lost. When a site disaster occurs, all data storage within a given geography is lost. This means that only data availability techniques that are geographically separated are effective against this type of data failure.

Even though data failures are relatively rare, they usually result in extended outage. The outage due to data failures is usually much longer than that caused by machine or process failure. Hours or even days of downtime are possible [2]. Applications that need high availability must therefore address data failures.

2   Physical and Logical Data Failures

We characterize data failures as either physical or logical based on where the data failure was introduced. We make this distinction because there are many solutions that only address physical data failures. The following figure shows how we model an application stack and the type of data failures.

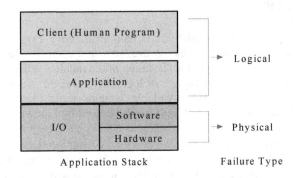

Fig. 1. Logical and Physical Data Failures

The I/O layer includes the hardware (adapter cards, cable, disk array) and software (file system, volume manager, and device driver). Applications such as databases use the I/O layer services. The client is a human or program that uses the services provided by the application layer.

A *physical data failure* is a data loss or data corruption that is introduced in the I/O software or hardware. A *logical data failure* is a data loss or data corruption that is introduced above the I/O software or hardware layers. We draw the boundary between physical and logical at the interface between the application and the I/O system because we want to describe the layering as seen from the application.

Note that a failure in a layer is not necessarily caused by code in that layer. A logical data corruption, for example, does not have to be caused by an application bug, it just has to be *introduced* while we are executing application layer code. For example, a memory fault that corrupts in-memory buffer used by the application would still cause a logical corruption even though the cause is hardware.

A logical corruption cannot be detected by I/O layer checks. This is because the data was already corrupted in the application layer by the time the data was passed to the I/O layer. The I/O system has no way to distinguish valid data from invalid data. More generally, a layer cannot detect errors introduced by layers above. An application, for example, cannot detect human errors. If a database administrator (client layer) were to issue a drop table command, the database (application layer) cannot always determine whether the human specified the correct table.

3 Physical Data Availability Techniques

In this section, we analyze technologies available for addressing physical data failures. Collectively, these techniques provide *physical data availability*. We start by discussing techniques for preventing physical corruptions, then techniques for recovering from physical data failure and site disaster.

Failure	Solution
Data Loss	RAID Backup/Restore Checksum Remote Snapshot
Data Corruption	Backup/Restore Snapshots Continuous Backup
Site Disaster	Offsite Backup Remote Mirroring

Fig. 2. Techniques for Physical Data Availability

3.1 Preventing Physical Data Failure

The primary means for preventing physical data loss is through redundant storage such as RAID [3]. Cyclic redundancy checksums (CRC) are also used to detect corruptions and error correcting code (ECC) memory is used to detect and repair errors [4]. These techniques have greatly improved the reliability of the I/O subsystem and reduced the frequency of data failures.

3.2 Recovering from Physical Data Failure

Backup/restore is the fundamental technique for recovering from physical data failures. When a data failure occurs, a backup of the data is restored. The limitations of using backups are data loss and restore time. After a backup has been restored, all the changes that were made since the backup was taken are lost. The exception is with applications such as databases that can use the transaction log to bring the backup to the current time. Restore time is an issue because a backup must first be restored from backup media before it can be used. Even with the use of on-disk backups instead of tape, restore time for large amounts of data can be lengthy.

To minimize data loss in the event of a restore, it is desirable to have frequent backups. Many storage vendors have optimized periodic backups so that they can be taken rapidly with relatively small disk overhead. These periodic backups are sometimes known as *snapshots* [5]. Snapshots can consist of full copies of the data (e.g., a mirror within the storage array), or a partial copy that tracks only changes since the last snapshot. Snapshots allow you to maintain multiple *restore points* to which you can bring the data in the case of a physical data failure. To address the issue of data loss after a restore, several *continuous backup* products [6,7] have recently been introduced. These apply database like logging techniques to track changes incrementally. This then allows you to return the data to any arbitrary point in the past.

3.3 Handling Site Disaster

Storing backup copies of the data off-site is a way to ensure that the data can be restored when the primary copy is unavailable. Remote disk mirroring can also be done so that one of the mirror copies is a remote storage unit. The remote copy is typically located in a data center that is not likely to be subject to the same disaster as the primary site. For example, it might be across a continent or an ocean. The geographic separation reduces the likelihood that a disaster that affects the primary site would also affect the backup site(s).

Note that remote mirroring over long distances can degrade the response time of the application because of propagation delays across long distances. In addition, for disaster recovery, all the other components of the application (application, hardware, networking) must also be available at the backup data center.

3.4 Limitations of Physical Data Protection Techniques

Physical data protection techniques provide protection against physical data failures. However, they only provide *partial* protection against logical data failures. In particular, physical data protection techniques can recover from, but cannot prevent, logical data failures.

3.4.1 Physical Techniques Cannot Prevent Logical Data Failures

Physical data protection techniques cannot *detect or prevent* logical data corruptions. This is because techniques such as RAID and checksums are physical checks. They cannot validate the contents of the data. So if a logically corrupted block were written to a RAID device, the storage cannot detect that the data is invalid. This limitation applies to all physical data protection techniques. For example, a storage array that is remotely mirrored would ensure that a logically corruption would propagate bit-for-bit to all remote mirrors. This is a serious limitation given the wide range of potential logical data failures.

3.4.2 Physical Techniques Cannot Prevent Upper Layer Physical Data Failures

Significantly, physical data protection techniques cannot even prevent physical corruptions that were introduced at a higher layer in the I/O system. Modern I/O subsystems have become very complex. There are usually many layers of software, firmware, and hardware that is involved in an I/O request. Some of these components include: file system buffer cache, volume manager, device driver, host bus adapter, storage area network switch, storage controller, and the hardware/software inside the storage unit. A bug or failure in any of these components can cause a physical corruption that would not be detected by lower layers of the I/O subsystem. For example, RAID cannot correct a directory entry that has been corrupted by a file system bug. From the perspective of the RAID device, the file system error is also a logical failure.

3.4.3 Physical Techniques Are Not Optimal for Recovering from Logical Data Failures

Physical data protection techniques such as backup/restore and snapshots can be applied to recover from logical data failure. So for example, if a file was overwritten or accidentally deleted, a copy can be restored from backup. Physical data recovery techniques, however, have many limitations. First, the recovery is done at the granularity of the physical object (file, file system, disk, etc.). If only a subset of the data were damaged, physical recovery cannot repair just the damaged part, it has to restore the entire object. So if a database file were a terabyte in size and only a single row was corrupted, a physical restore would bring back the entire terabyte; it cannot restore just the row.

Doing file- or device-based recovery is not optimal because recovery takes longer and you will lose more data. Recovery takes longer since you are usually restoring more data than you need. Restoring an object typically means loss of data because the backup does not contain changes made since the backup was taken. Because the granularity is at the physical file or device, the data loss is also larger than required. For example, restoring the entire file means that you will lose changes made to all the records in the file even though only a single record is damaged.

In summary, physical data protection techniques are inadequate for applications that require high levels of data availability.

4 Logical Data Availability Techniques

We now look at how logical data protection techniques can address the limitations associated with physical data protection techniques. Logical data protection techniques exploit application knowledge to offer better protection against data failures and better recovery from data failures. We will use the Oracle database as an example and show how it exploits knowledge of the logical structure of data to provide improved data availability. Figure 3 shows some of the solutions that we'll cover in this section.

Objective	Solution
Preventing Logical Data Failure	End-to-End Application Checksum Standby Database
Data Corruption	Backup/Recovery Flashback Precision Repair

Fig. 3. Techniques for Logical Data Availability

4.1 Preventing Logical Data Failure

The main logical data failure prevention techniques are application level checksums and logical replication.

4.1.1 End-to-End Application Checksum

Physical data protection technologies such as RAID and checksums are limited because they only validate the data when it is in the I/O subsystem. These techniques are ineffective against data that were corrupted prior to the initiation of the I/O request.

Applications can achieve higher data availability by implementing their own application level checksums. For example, when Oracle writes data, it performs a series of logical and physical checks on the data to ensure integrity. As part of this, it also stores checksum and other validation information in the data blocks. When the data is read back, Oracle verifies the checksum and other validation information in the blocks. If the block fails the checks, an error is raised. This detects I/O corruptions and prevents bad data from being used. Microsoft Exchange Server has a similar capability [8].

Conventional application level checksums can detect a corruption after the fact. Oracle's HARD (Hardware Assisted Resilient Data) [9] takes this further by preventing I/O corruptions from making it to disk in the first place. Figure 4 shows how this technology extends the range of validation all the way from the application (the database) to the hardware.

Under the HARD initiative, Oracle worked with storage vendors to imbed knowledge of Oracle block formats and checksums within storage arrays. When Oracle data is written to a HARD-compliant storage device, the storage device can independently validate the integrity of the Oracle blocks as well as the locations to which the blocks were destined. If the validation checks pass, the data is written to the disks. If the validation checks fail, the write is not performed and an error is returned. This ensures that corrupted data are never written to disk.

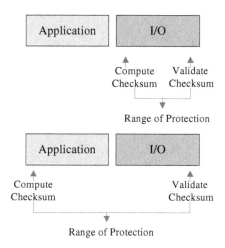

Fig. 4. End-to-end Application Checksum

The end-to-end application checksum can prevent a whole class of data failures. It protects against data corruptions introduced by any subsystem between the application (Oracle) and the disk array, it prevents misdirected writes, it also prevents overwrites of Oracle files by other applications.

The end-to-end application checksum also protects data during direct disk-to-tape transfers. Certain disk arrays can directly transfer data to tape drives without host intervention. This is useful for high performance backups. A corruption in this case is very severe as it means that the backups that you'll need to recover from other data failures would be corrupted. If the tape device is HARD compliant, it can directly validate the integrity of the data that it is receiving.

4.1.2 Standby Database

Application-maintained replicas, such as Oracle Data Guard [10,11] and Sybase Replication Server [12], are another powerful means for protecting data against both physical and logical failures. A standby database is a copy of the database that is kept up-to-date with changes as they are made on the primary database. Figure 5 shows how it works in a 2 database configuration:

1. A change is made on the primary database
2. The change is captured in the redo log
3. The log is shipped to the standby database
4. The change described in the log is applied to the standby database

If the primary database fails, the standby database can become the primary and continue operation. A primary may have multiple standby databases, each of which can be at a different location. A standby database that is geographically remote from the primary database can also offer protection against site failure.

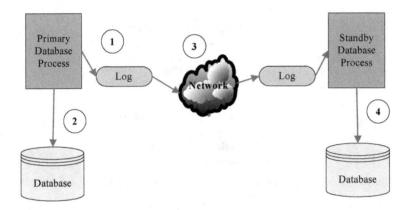

Fig. 5. Standby Database Operation

A standby database can be thought of as a remote mirror. As such, it can protect against the same physical data failures as storage-level remote mirroring. A standby, however, is more resilient against logical corruptions. This is because of how changes are propagated and applied at the standby databases. A storage-level remote mirroring solution sends description of the physical blocks that are changed. So if a logically corrupted block were written to the primary mirror, the remote mirroring software/firmware will ensure that the corruption is replicated bit-for-bit at the remote mirror.

Oracle Data Guard, on the other hand, receives a logical description of the change from the primary database. The redo log contains a great deal of contextual information such that if the log is corrupted or if the blocks being modified are inconsistent with the redo log record, the change application (described in step 4 above) would fail. This prevents many logical corruptions from propagating and corrupting the copy of the data on the standby.

4.1.3 Improving the Effectiveness of a Standby Using an Apply Lag
One can improve a standby database's effectiveness by building in an *apply lag*. In this mode, changes are still shipped immediately to the standby; however, the changes are not applied until after a configuration delay. Note that the lag does not mean that the standby will lose data after a failover. This is because there is no delay on receiving the changes at the standby, just on when these changes are applied. Since all the changes are available on the standby database, no data would be lost if the primary database were to fail.

The lag gives the customer a chance to react to human error or application corruption before the standby database is affected. For example, if one accidentally deleted (dropped) a table on the primary database, the delay means that the operation would not be immediately carried out at the standby. This delay gives the system a chance to discover the error and immediately stop the standby apply. You can then apply the changes up to the point of the error and still have a good copy of the database.

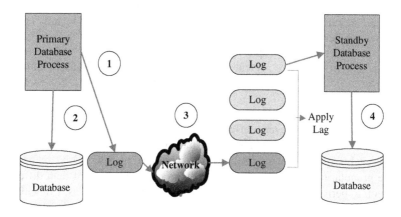

Fig. 6. Using Standby with Apply Lag

The use of a lag comes with a trade-off in safety vs. failover time. For maximum data protection, customers like to configure a long apply lag. This gives them more time to detect and respond to a data failure on the primary before the same data failure also makes it to the standby. But a long apply lag means that there would be more log data on the standby database that are not yet applied. When the standby takes over from a primary database after the primary has failed, the standby needs to apply all the unapplied logs before accepting new transactions. This increases the failover time. Another problem with the apply lag is that queries on the standby database do not see current data.

One way customers have addressed this problem is to have multiple standby databases each configured with a different lag. For example, a customer might have several standby databases that are 5 minutes, 30 minutes, 2 hours, and 1 day behind the primary. When the primary encounters a data failure, they activate the standby with the least lag that doesn't have the same corruption. This is analogous to the way some customers keep multiple snapshots of their data. The difference is that all the standby 'snapshots' are continuously brought up to date (within the constraint of the apply lag).

Oracle Data Guard does not have this trade-off because of the *flashback database* feature. Flashback database [13,14] allows one to rapidly undo recent changes that were made. It does so by undoing all the changes in reverse order. With flashback database enabled, a standby database would immediately apply the changes as they are shipped from the primary database. This way, the standby database is current as of the last change and incurs no delay during a failover. In case of a human error or a logical corruption on the primary (conditions that previously were handled by the apply lag), the standby would simply flashback the database to a point before the error took place.

4.2   Recovery from Logical Data Failure

End-to-end application checksum validation such as HARD and application-maintained replicas such as the standby database can prevent data failure due to human error and logical corruptions. Nevertheless, there are still situations in which one needs to recover from a logical data failure. This section discusses some of these techniques.

4.2.1   Logical Backup and Recovery

Logical backup and recovery offers finer granularity than physical backup and recovery. For example, database oriented backup and recovery can deal with entire databases, tablespaces, tables, database blocks, and even rows. For backups, control over the granularity means that you can backup different types of data differently. It allows you to perform backups faster and create smaller backup sets. Logical recovery is where application knowledge becomes really important; it can significantly reduce downtime. With Oracle Recovery Manager [15], for example, if only a few blocks in a database are corrupted, you can restore just the damaged blocks and then apply the logs to bring those blocks current instead of restoring and recovering an entire datafile.

Logical recovery also provides much finer control over the point of time to which to recover the data. A database, for example, can use its transaction logs to re-apply the changes made since time of the backup or snapshot. This allows databases to recover to any point in time, given a backup as a starting point. Thus most recoveries would not lose data.

4.2.2   Application-Level Continuous Backups

Oracle has similarly augmented its backup and recovery capabilities with flashback technology. Flashback database is like a continuous backup for the database. To recover the database to a prior point in time, flashback database replays the log in reverse, therefore undoing the operations in reverse order. Unlike conventional recovery, there is no need to restore a backup first. Consequently, flashback is extremely fast when the objective is to undo a mistake or corruption in the recent past.

Flashback database is more efficient than physical continuous backups because flashback database can take advantage of the regular database logs. Like physical restore points, flashback database can also have named points to which you can bring the database back. Because these restore points simply name points in the existing log stream, flashback database restore points do not consume as much resources as conventional split mirror restore points.

Flashback database is a database level continuous backup capability. Some research work has also been done on undoing data failures in mail servers [16].

4.2.3 Precision Data Repair

Application-level data repair can achieve extremely fine-grained data repair. A good example of using application logic to perform fine-grained data repair is an Oracle feature called *flashback query* [13]. Flashback query takes advantage of the Oracle database's multi-version consistency scheme in which the system maintains metadata so that it can reconstruct versions of data in the past. The use of flashback query is best illustrated by an example:

Suppose someone erroneously deleted all the rows corresponding to people who reported to the manager named Jon Smith. That information is now gone from the database.

To get a list of the people that existed at that point in time, one can issue a query similar to the following:

```
SELECT * FROM employee AS OF TIMESTAMP
   TO_TIMESTAMP('2004-12-04 02:45:00',
                'YYYY-MM-DD HH:MI:SS')
   WHERE manager = 'JON.SMITH';
```

This would return a list of employee that reported to Jon Smith at that point in time. Using this information, the data can be reinserted into the database.

Flashback query therefore allows you to find the exact version of the data that you want. Once this is found, it can be reinserted into the table. Since it is expressible as SQL, you can use the full query capabilities to identify the data that was lost and reinsert them.

Along with flashback query, Oracle also supports:

- *Flashback version query* that gives you all the versions of a row between two times and the transactions that modified the row
- *Flashback transaction query* that allows you to see all the changes that were made by a transaction

The combination of the various flashback capabilities makes it practical to surgically repair corrupted data rapidly with minimal data loss. These capabilities cannot be achieved by physical data recovery techniques.

5 Conclusion

This paper makes the argument that logical data protection and recovery techniques are more powerful and offer higher data availability than traditional techniques that work at the physical level. Logical data protection and recovery accomplish this by exploiting application-level knowledge. Physical data protection cannot provide this higher level of data availability. For example, RAID can protect against disk failures but cannot prevent logical corruption or human error. Physical data recovery techniques also have shortcomings in terms of recovery time and data loss.

Fortunately, every physical data protection technique has logical counterpart. Figure 7 summarizes the corresponding physical and logical techniques.

	Physical	Logical
Data Protection	RAID Checksum	HARD (End-to-End Application Checksum)
Data Recovery	Backup/Restore Snapshots Continuous Backup	Backup/Recovery to Arbitrary Points Precision Data Repair Restore Point Flashback Database
Site Disaster	Offsite Backup/ Restore Remote Mirrors	Offsite Backup/Recovery Standby Database

Fig. 7. Physical and Logical Data Protection Techniques

Each of the logical data availability techniques listed is more powerful than its corresponding physical data availability technique. By taking advantage of application knowledge, logical data availability techniques can detect and prevent more errors and offer more and finer-grained data recovery options.

There is still a place for physical data protection techniques. The Oracle database is exceptional in the breadth of logical data protection techniques that it supports. Most applications do not have similar capabilities. These applications must therefore rely on physical data protection techniques. Because physical data protection techniques do not have application knowledge, they can support all applications. Even though it is not as effective as logical data protection, physical data protection is the only choice when there are no logical data protection techniques for the application under discussion.

References

1. Donna Scott, *Continuous Application Availability: Pipe Dream or Reality*, Gartner Data Center 2003, (Las Vegas, NV), page 9, 8-10 December 2003.
2. Tim Wilson, *Ebay retrenches: Devastating outage exposes lack of redundancy, need for simplicity*, InternetWeek.com, June 1999.
   Available at http://www.internetweek.com/lead/lead061799.htm.
3. David A. Patterson, Peter Chen, Garth Gibson, and Randy H. Katz. *Introduction to redundant arrays of inexpensive disks* (RAID). Spring COMPCON'89 (San Francisco, CA), pages 112-17. IEEE, March 1989.
4. W. Wesley Peterson and E.J. Weldon, Jr., *Error-Correcting Codes*, 2nd edition, MIT Press: Cambridge, Mass., 1972.
5. Dave Hitz, James Lau, Michael Malcolm, *File System Design for an NFS File Server Appliance*, Proceedings of the USENIX Winter 1994 Technical Conference, January 1994.
6. Evan Koblentz, *Continuous Backup*, eWeek, June 23, 2003.
7. Symantec's Nortan GoBack 4.0 data sheet.
   Available at: http://www.symantec.com/goback.

8. *Understanding –1018 Errors.* June 2001. Available at: www.microsft.com/technet.

9. *Oracle Hardware Assisted Resilient Data (HARD) Initiative,* 2004, available at http://www.oracle.com/technology/deploy/availability/htdocs/HARD.html

10. *Oracle Data Guard in Oracle Database 10g – Disaster Recovery for the Enterprise,* December 2003, available at: http://www.oracle.com/technology/deploy/availability.

11. *Oracle Data Guard Concepts and Administration 10g Release 1 (10.1),* Part Number B10823-01, December 2003, Oracle Corporation.

12. *Replication Server 12.6 Features,* 2003, available at: http://www.sybase.com.

13. *Flashback Technology,* 2004, available at:
   Available at: http://www.oracle.com/technology/deploy/availability.

14. Oracle Database Concepts 10g Release 1 (10.1), Part Number B10743-01, December 2003, Oracle Corporation.

15. Oracle Database Backup and Recovery Basics 10g Release 1 (10.1), Part Number B10735-01, December 2003, Oracle Corporation.

16. Aaron B. Brown and David A. Patterson, *Undo for operators: Building an Undoable E-mail Store,* Proceedings USENIX Annual Technical Conference, San Antonio, TX, 2003.

Contract-Based Web Service Composition Framework with Correctness Guarantees

Nikola Milanovic

Humboldt University, Berlin
milanovi@informatik.hu-berlin.de

Abstract. We present formal and practical foundations for Web service composition framework with composition correctness guarantees. We introduce contractual composition model based on two isomorphic description models: Contract Definition Language (XML) and abstract machines (formal notation). Composition operators (patterns) are used to perform composition which is then formally verified with respect to properties described in service contracts. We also describe Java-based implementation of the system, concentrated around Sun's Java Web Services Development Pack (JWSDP).

Indexed Terms: Web services, composition, correctness, contracts.

1 Introduction

Web services are emerging as a replacement and/or additional paradigm for the component-based software development. However, Web services aim much further to become not only a new Object Request Broker architecture, but a unifying paradigm for communication among heterogenous groups of software and hardware entities. Web service architecture has three layers: description and basic operations (publication, discovery, selection and binding), composite services (coordination, conformance, monitoring and quality of service) and managed services (certification, rating, liability). Unfortunately, only the bottom layer has been standardized (WSDL, UDDI and SOAP). We are seeking a solution for the second layer dealing with Web service composition. In the next section we discuss the ideas of existing approaches and then present formal foundations and implementation of a contract-based composition framework.

2 Related Work

The composition layer comprises four properties: coordination, conformance, monitoring and quality of service. Coordination determines which services participate in composition and how they exchange messages. Conformance establishes composition correctness, while monitoring basically deals with error and exception handling. Finally, quality of service offers metrics to compare different compositions with respect to nonfunctional properties.

M. Malek, E. Nette, and N. Suri (Eds.): ISAS 2005, LNCS 3694, pp. 52–67, 2005.

We examined several approaches for Web service composition: Business Process Execution Language for Web Services (BPEL4WS) [5], Semantic Web (OWL-S) [6,11,17], Web component [19], π-calculus [12], Petri Nets [10], model checking [9] and finite state machines [3,4]. Our detailed survey of these solutions and how they relate to the four key composition properties can be found in [15].

The main problem with 'industrial' approaches (BPEL, OWL-S) is the lack of support for verifying composition correctness. Both approaches (and BPEL in particular) offer implementation languages that are simply too expressive for any kind of formal validation. On the other hand, there are other, more formal and abstract approaches (e.g., π-calculus or finite state machines). However, they are often difficult to apply in real-world scenarios, and some of them face serious scalability problems. Our intention is to provide formal and practical foundations for a contract-based composition approach with correctness guarantees.

3   Contracts and Abstract Machines

The concept of design by contract was first introduced in [13] to facilitate component reuse. A contract describes, in a standard way, what a component expects from its clients and what it delivers if those requirements are met. We propose to use contracts as non-functional (QoS) extension of WSDL description.

3.1   Contract Definition Language (CDL)

We assume that messages and port bindings are already defined, separate (WSDL) or as a part of the service contract. The contract itself must provide information that is related to non-functional aspects of the service execution. However it should not include implementation details but semantic information only: what a service will deliver and under which conditions it will execute correctly.

The root section of CDL schema (Figure 1a) comprises `organization`, `types`, `location`, `method` and `event` elements, as well as several basic attributes: uri, name, description, price, state information, version and port. The `organization` element is introduced to maintain backwards compatibility with Universal Description, Discovery and Integration (UDDI) directories. Therefore, every service must belong to an organization that publishes it. For each organization it is possible to define name, description (keywords), classification and primary contact. The `types` element describes complex types that a service supports. It is used when a service accepts or returns non-primitive types (e.g., object of come custom class), and clients should be able to construct/deconstruct them appropriately. The `location` element is introduced to support location-based services. It allows for definition of country, city, street address and GPS coordinates. The `event` element declares all events that a service supports. For each event, its native name is listed, along with a reference and a common environment exception it is mapped into (if available). Finally, the `method` element describes one or more service methods.

Inside the `method` element we can specify information about parameters, persistent resources, invocation, pre-conditions, post-conditions and invariants, events, assertions, classification and method location. For each parameter it is possible to define name, type, restriction and initialization. Furthermore, constants and sets (complex types) that a method understands can be listed. Invocation information is related to component creation and execution (synchronous or asynchronous). Pre-conditions, post-condition and invariants share the same structure (Figure 1b).

Pre-conditions are linked to exported methods and determine obligations of a client. A method is guaranteed to work correctly if and only if a client satisfies pre-condition. Post-condition describes what a method guarantees, if precondition holds. Invariants are properties that must hold before and after invocation of each exported method. They describe general, static properties of a service. The properties that can be described are: rendering, logging, security, dependability (transactions, replication, check-pointing, timeout and exceptions), performance and parameters.

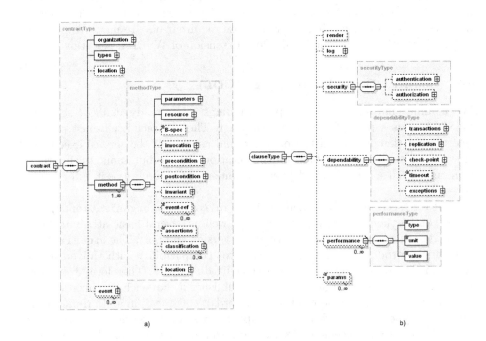

Fig. 1. The root contract structure

We shoved in [16] that it is possible, to some extent, to automate extraction of defined properties from Java classes and Enterprise Java Beans. We also identified the benefits that early contract inclusion has on the typical software lifecycle.

3.2 Modeling Services as Abstract Machines

Our main goal is to support not only reuse, but composition of Web services. CDL syntax offers a richer set of description primitives compared to WSDL, that can be used for specifying relevant non-functional properties. Verification of composition correctness, however, requires a formal approach.

Since contracts, as we presented them so far, are just plain text (XML) files, it would be very difficult, if indeed possible, to judge correctness of their composition. The first problem we would be faced with is actual definition of correctness. What does it mean for a contract to be correct, apart from satisfying XML requirements of being well-defined and well-formed? How can we judge whether two or more contracts are compatible or not conflicting with each other? How to define relations "compatible" and "conflicting"? Finally, how to perform actual composition when working on text files? In order to be able to answer these questions, we introduce a second form for expressing Web service contracts: abstract machine notation (AMN). We need the XML notation in order to transport contracts over a network (interoperability), while AMN serves the purpose of giving contract elements formal mathematical treatment.

Abstract machines are specified using Abstract Machine Notation. Details on AMN can be found in [1]. We give a short overview of AMN principles. An element, which can be a class, a component, or a Web service, is represented as an *abstract machine*. It is characterized by *statics* and *dynamics*. The statics corresponds to the definition of the state, while the dynamics corresponds to the operations:

```
MACHINE M(X,x)
CONSTRAINTS C
CONSTANTS c
SETS S; T={a,b}
PROPERTIES P
COMPLEX Cx
VARIABLES v
INVARIANT I
ASSERTIONS J
INITIALIZATION U
OPERATIONS
 u1 <- O1(w1) = PRE Q1 THEN V1 END
 ...
 un <- On(wn) = PRE Qn THEN Vn END
END
```

This is a parameterized abstract machine having free dimensions X (set) and x (scalar). CONSTRAINTS describes conditions on machine parameters. SETS contains finite or named sets that the machine can use, while CONSTANTS describes constants that the machine understands. PROPERTIES takes form of conjoined predicates specifying invariants involving constants and sets. VARIABLES lists state variables, and INVARIANT describes static properties of the machine, that must be preserved before and after each operation. ASSERTIONS is deducible from PROPERTIES and INVARIANT, and exists purely to ease the proving of machine correctness. INITIALIZATION initializes state variables. OPERATIONS lists operations of an abstract machine, with pre-conditions (PRE) and post-conditions (THEN).

Operation body of an abstract machine modifies a machine state. For expressing formally how such modification takes place, we will be using logical predicates relating the values of state variables just before the operation is invoked to the values just after the operation completes. The method we use is called substitution. Let P be a formula, x be a variable and E an expression, then the following denotes the formula obtained by replacing (substituting) all free occurrences of x in P by E:

$$[x := E]P$$

If S is a substitution, and I is a formula, we write that substitution S preserves I in a following way:

$$I \implies [S]I$$

This expression says that if the invariant I holds then the substitution S is guaranteed to preserve the same invariant. We now introduce more complex substitutions.

- Pre-conditioned substitution: If P is a pre-condition and S is the substitution guarded by this pre-condition, then pre-conditioned substitution is $[P|S]R \iff P \wedge [S]R$. This substitution can also be noted as PRE P THEN S END.
- Multiple simple substitution: Often we have to perform simultaneous substitution (multiple simple substitution) $[x, y := E, F]P \iff [x := E][y := F]P$. It can be expressed also as x:=E || y:=F.
- Bounded choice substitution: It is used when we have to express a choice between two or more substitutions. It is denoted with $[S \Box T]R \iff [S]R \wedge [T]R$, and can also be expressed with CHOICE S OR T END.
- Guarded substitution: A substitution can be guarded by a predicate using implication $[P \implies S]R \iff (P \implies [S]R)$. It can be denoted IF P THEN S END.
- Conditional substitution: Combination of bounded choice and guarded substitution is called conditional and is defined IF P THEN S ELSE T END $\iff$ $(P \implies S) \Box (\neg P \implies T)$.
- Empty substitution: A substitution that performs nothing for the target post-condition is empty substitution $[skip]R \iff R$.
- Multiple generalized substitution: Any combination of previously defined substitutions can be performed simultanously (multiple generalized substitution) using the same notation, e.g., $S||(P|T) = P|(S||T)$, $S||(T \Box U) = (S||T) \Box (S||U)$, $S||(P \implies T) = P \implies (S||T)$.
- While substitution: A situation where, if a predicate P holds, substitution S is iteratively executed, is denoted with WHILE P DO S END.

Mapping from CDL to AMN. We have presented two notations for describing Web service contracts. During service exploitation there will be times when we will have to switch between them:

- When composing two services, their CDL descriptions will be transferred into abstract machines to allow for formal treatment of their properties.
- When new service is composed it is constructed by merging abstract machines of the constituent services, thus producing another abstract machine. In order to make this service available to others and to be able to transport its specification over a network, abstract machine has to be transferred into CDL description.

It can be seen that transformation between CDL and AMN has to be bidirectional. However, since this transformation is linear, once we know how to do it one way, the other way is trivial. The mapping algorithm works as follows:

1. Machine name is constructed from `serviceName` attribute of the `contract` element. All other attributes of the `contract` element, as well as all child elements and attributes of the `organization` and `location` elements are mapped into `CONSTANTS` clause.
2. The `types` element is mapped into `COMPLEX` clause of abstract machine.
3. The `event` element is mapped into `CONSTANTS` clause.
4. For each `method` element, the following is performed:
   (a) State variables are built from properties in `invariant`, `precondition`, `post- condition` elements. To this are added all method parameters.
   (b) All sets defined in the `set` element are added to the `SET` clause.
   (c) Constants from the `constants` element are added to the `CONSTANTS` clause.
   (d) The `INVARIANT` clause is defined in term of conjoined predicates involving state variables, and is mapped directly from the `invariant` element. The `INVARIANT` clause must contain enough conjuncts to allow for the typing of all state variables.
   (e) The `PRE` clause is mapped directly from `precondition` element. State variables designating input parameters must have constraints (or types) defined in this clause.
   (f) Operation body (postcondition, or `THEN` clause) is constructed by conjoining substitutions from the `postcondition` element. All output parameters must have properties (or types) described in this clause.
   (g) All state variables that have `initialization` element defined, are added to the `INITIALIZATION` clause. Additionally, those that are defined as `"INOUT"` are added to the list of machine formal parameters.
   (h) The content of `assertion` element (if exists) is added to the `ASSERTION` clause in form of conjoined predicates.
   (i) The `resource`, `invocation` and `event-ref` elements are of no interest for composition semantics, and are thus not transferred into AMN. They are used for maintaining internal consistency of composition process.

4   Service Composition

We identify four basic patterns (operators) for service composition: sequence, choice, parallel and loop. We show how to construct composite abstract machine clauses for each case and then discuss verification of composition correctness.

4.1   Sequential Composition

The sequence operator (Figure 2a) executes two (or more) services in an ordered sequence. We denote sequential composition of services A and B with $C = A \triangleright B$. Outputs of the left operand (A) become inputs of the right operand (B). The clauses of the resulting abstract machine are calculated:

- SETS, CONSTANTS, and VARIABLES clauses are concatenated
- PROPERTIES, INVARIANT, and ASSERTION clauses are conjuncted
- OPERATIONS clause is constructed by performing substitution of the left operand, then substituting input state variables of the right operand with the output state variables of the left operand, then performing substitution of the right operand, while conjuncting preconditions:

$$\text{OPERATION } output_B \leftarrow C(input_A)$$
$$\text{PRE}\quad P_A \wedge P_B$$
$$\text{THEN}\quad S_A \text{ ; } input_B := output_A \text{ ; } S_B\quad \text{END}$$

Here $output_B$ is a set of output state variables of the right operand, $input_A$ is a set of input state variables of the left operand, C is the name of a new (composite) operation, $input_B$ is a set of input state variables of the right operand, and $output_A$ is the set of output state variables of the left operand.

- INITIALIZATION clauses are concatenated, and multiple composed if needed.

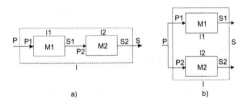

a)   b)

Fig. 2. Sequence and Choice Patterns

4.2   Parallel Composition

Parallel composition executes two (or more) services concurrently. We allow two subtypes of this pattern: parallel composition with communication (Figure 3a) and without communication (Figure 3b). In the former case, concurrent services can communicate with each other, for the purpose of synchronization of some state variables. It can be used when a certain decision has to be reached after parallel processing has been performed, e.g., choosing result of one service and discarding the other. Only operators of the relational algebra are allowed for the state variables upon which the synchronization is performed. We do not allow any kind of result aggregation, since it would needlessly complicate composition pattern. If data aggregation needs to be performed, additional service should be created and then sequentially composed to the parallel composition. In the latter case (no communication), there is no communication / synchronization between concurrent services.

Parallel composition with communication is denoted with $\|_{P(c)}$, where c are state variables that are being used for synchronization and P is the predicate evaluated upon them, while parallel composition without communication is denoted with $\|$: $C = A \|_{P(c)} B$ and $C = A \| B$. The clauses of the composed abstract machine are constructed:

- SETS, CONSTANTS, and VARIABLES clauses are concatenated
- PROPERTIES, INVARIANT, and ASSERTION clauses are conjuncted
- OPERATIONS clause is constructed differently for composition with and without communication:
  - For parallel composition without communication, pre-conditions are conjuncted and substitutions are performed simultaneously (using multiple general substitution):

    OPERATION $output_C \leftarrow$ C($input_C$)
    PRE $P_A \wedge P_B$
    THEN $S_A \| S_B$ END

    Here $output_C = output_A \cup output_B$ and $input_C = input_A \cup input_B$.
  - For parallel composition with communication, pre-conditions are conjuncted and substitutions are performed simultaneously. Afterwards, predicate P is evaluated on a subset of state variables c, resulting in choice of output of only one service:

    OPERATION $output_C \leftarrow$ C($input_C$)
    PRE $P_A \wedge P_B$
    THEN $S_A \| S_B$

    IF P_c THEN $output_C = output_A$ ELSE $output_C = output_B$ END
- INITIALIZATION clauses are concatenated, and multiple composed if needed.

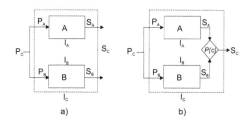

Fig. 3. Parallel Pattern

4.3   Choice Composition

The choice pattern (Figure 2b) represents a composition that behaves as either of its constituents services. It is similar to parallel composition pattern with communication, but it is non-deterministic. It is furthermore restricted to compatible services in sense of input parameters, because it is used when it is known in advance that some of the available services can perform the requested operation, without the need to know which one will do so in a particular instance. The most general example is sending the same request to many services and accepting the results from the one that first completes its execution. This composition pattern is denoted with $C = A \odot B$. The machine resulting from applying choice pattern is constructed as follows:

- SETS, CONSTANTS, and VARIABLES clauses are concatenated
- PROPERTIES, INVARIANT, and ASSERTION clauses are conjuncted
- OPERATIONS clause is constructed by conjoining preconditions and connecting substitutions by bounded choice substitution operator:

$$\text{OPERATION } output_C \leftarrow \text{C}(input_C)$$
$$\text{PRE} \quad P_A \wedge P_B$$
$$\text{THEN} \quad S_A \square S_B \text{ END}$$

Here $output_C = output_A \vee output_B$, which is implied in $S_A \square S_B$.
- INITIALIZATION clauses are concatenated, and multiple composed if needed.

4.4   Looping

Looping pattern supports execution of the same service repeatedly, until a certain condition is fulfilled. Based on the condition controlling the loop, we define unary (Figure 4a) and binary loop (Figure 4b): $C = \circlearrowleft_{P(e)} A(e)$ and $C = W(e) \circlearrowleft_{P(e)} A$. In both cases, looping is controlled by predicate P evaluated on the variable e. Service is executed until $P(e)$ becomes false. In the unary pattern, e is a state variable of service A, and is changed in every iteration by the execution of A. Therefore, service A controls the loop exit condition. Since this is the loop with the condition on top (exit condition is evaluated prior to execution), variable e must be in the INITIALIZATION clause to enable the first loop iteration. In the binary pattern, there is another service W that controls $P(e)$. In this case we do not allow service A to influence the loop exit condition. Here, service W is executed prior to A and will set value of e, which therefore does not have to be initialized. The composite machine is constructed as follows for the unary pattern:

- The clauses SETS, CONSTANTS, VARIABLES, PROPERTIES, INVARIANT, ASSERTION, and INITIALIZATION are kept unchanged. Variable controlling loop exit (e) must appear in the INITIALIZATION clause.
- Operation body is constructed by enclosing original substitution in a WHILE DO block, controlled by $P(e)$:

$$\text{OPERATION } output_C \leftarrow \text{C}(input_C)$$
$$\text{PRE } P_A$$
$$\text{THEN WHILE } P(e) \; S_A(e) \; \text{END}$$

Here $ouptut_C = output_A$ and $input_C = input_A$.

For the binary pattern, another service W controls exit variable:

- SETS, CONSTANTS, and VARIABLES clauses are concatenated
- PROPERTIES, INVARIANT, and ASSERTION clauses are conjuncted
- Operation body is constructed by conjoining preconditions, and enclosing both substitutions inside a WHILE DO:

$$\text{OPERATION } output_C \leftarrow \text{C}(input_C)$$
$$\text{PRE } \quad P_A \wedge P_W$$
$$\text{THEN } S_W(e)$$
$$\text{WHILE } P_e$$
$$S_A; S_W(e) \; \text{END}$$

- INITIALIZATION clauses are concatenated, and multiple composed if needed.

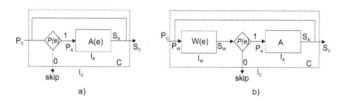

Fig. 4. Loop Pattern

4.5   Correctness Verification

Once an operator has been applied, composition result has to be verified. The whole composition process then proceeds as follows:

- Merging of two (or more) abstract machines using composition operator.
- Type checking of the resulting abstract machine.
- Proving correctness of the resulting abstract machine
- Establishing correct termination of the resulting abstract machine

Type Checking. Suppose that we have an expression E and a set s such that $E \in s$. Suppose further that there exists a set t such that $s \subseteq t$. Then it follows that $E \in t$. We can continue by including t in a larger set u, and then E will also belong to u. The purpose of type-checking of the abstract machine is to provide an upper limit for such set containment for all predicates. This upper limit is called a super-set of s and is at the same time the *type* of E. The function check is introduced like ENV $\vdash$ check (P), and for the predicate P means that within the environment (antecedent) ENV predicate P type-checks. Referring to the abstract machine from Section 3.2, the type checking consists of the following requirements:

- X, x, S, T, a, b, c, v are all distinct
- Operation names of $O_1...O_n$ are all distinct
- S, T, a, b, c, v are not-free in C
- v, X, x are not-free in P
- $X, S, T, a \in T, b \in T \vdash$ check $(\forall x(C \Rightarrow \forall c(P \Rightarrow \forall v(I \wedge J \Rightarrow U \wedge O))))$

The last expression means that first universally quantified scalar parameters and their constraints are checked, then universally quantified constants and their properties, then universally quantified variables and their invariant, and finally initialization and operations.

Proof Obligation. After type checking has been performed, the resulting machine must be proved correct. The purpose of this is to establish the following:

- Composite initialization must establish composite invariant
- Composite assertion must be deducible from composite properties and invariant
- Composite operation must establish composite invariant

Formally, and again referring to the machine from Section 3.2 we can write it:

$$C \wedge P \Rightarrow [U]I$$

$$C \wedge P \wedge I \Rightarrow J$$

$$C \wedge P \wedge I \wedge J \wedge Q \Rightarrow [V]I$$

Correct Termination. After proving machine correct, we have to see whether it will terminate correctly and whether it is feasible. For a given substitution S the construct $trm(S)$ denotes the predicate that holds when substitution S terminates, that is, establishes its post-condition. By requiring that all operations terminate, we ensure elimination of deadlocks. Another construct is defined, $abt(S)$ which denotes aborted substitution, that is, substitution that does not establish anything. Therefore it can be said that $abt(S) \equiv \neg[S]R$ for any predicate R, and accordingly $trm(S) \equiv \neg \, abt(S)$. We define correct termination as $trm(S) \iff [S](x = x)$. Correct termination for substitutions is established in the following way: $trm(P|S) \iff P \wedge trm(S)$, $trm(P \Box T) \iff trm(S) \wedge trm(T)$, $trm(P \Rightarrow S) \iff P \Rightarrow trm(S)$.

We also check whether the composite operation is feasible, with respect to the guarded substitution. The feasible operation will establish one, or none post-condition. Non-feasible operation, on the other side, will be able to establish any post-condition. We define feasibility as $fis(S) \iff \neg[S](x \neq x)$. The feasibility of the standard substitutions is calculated in the following way: $fis(P|S) \Longleftrightarrow P \Rightarrow fis(S)$, $fis(P\square T) \Longleftrightarrow fis(S) \vee fis(T)$, $fis(P \Rightarrow S) \Longleftrightarrow P \wedge fis(S)$.

Correct termination for loop operator is outside the scope of this general paper. Suffice to say that we introduce two separate elements in the loop body: invariant and variant (exit condition). We currently limit the evaluation of the exit condition to natural number and observe its monotonicity:

$$trm(\texttt{WHILE P DO S INVARIANT I VARIANT V END}) \Longleftrightarrow$$

$$(\forall x \cdot (I \Rightarrow V \in N)) \wedge (\forall x \cdot (I \wedge P \Rightarrow [n := V][S](V < n)))$$

5 Implementation

In previous sections we formed formal foundations needed for developing Web service composition framework. Now we address some implementation issues, namely system organization, communication, service directory and state management. Composition engine comprises four main parts: client application, basic administrative services, database for storing CDL contracts, one or more application servers/containers with deployed services. Service descriptions are stored in a relational database. Client applications access basic functionalities of the engine via Web service middle layer. This middle layer connects to the underlying database, as well as to application servers in which target services are deployed. Clients cannot access database or application servers directly. Therefore, most of the engine's tasks are accomplished in the middle layer. The engine is implemented using Java and Java-related technologies.

5.1 Client and Basic Administrative Services

Client part is realized as Swing application connected to the middle Web service layer that provides administrative functions and operations. Middle layer offers the following operations: publishing new service to directory, modifying and deleting existing service from directory, searching for services, composing new services using existing ones, invoking single or composed services.

In order to achieve these tasks, middle layer communicates with underlying relational database (directory) and application servers hosting target Web services that users want to invoke and/or compose. Since entire application is Web service-based, the communication is realized using Sun's Java Web Services Developer Pack (Sun JWSDP) [14].

We found two technologies provided within JWSDP very useful: Java Architecture for XML Binding (JAXB) and Java API for XML-based Remote Procedure Calls (JAX-RPC). Since CDL schema is very large, encompassing more than 50 complex entities, we need a powerful yet flexible mechanism of translating XML document into Java object representation. JAXB offers a complete solution for transferring XML content into Java object representation and vice versa. JAXB operation is based on three actions: binding XML schema to Java content classes, unmarshalling XML document into content classes and marshalling content classes into XML document. At the beginning CDL schema is compiled with JAXB binding compiler. This action produces a set of Java content classes that reflect the contract structure. The process of unmarshalling takes service contract as input and produces set of instantiated Java content classes populated with data parsed from XML document. During unmarshalling contract is optionally validated with respect to schema. After this step we have in-memory representation of contract. The middle layer uses JAXB to publish and modify service contracts. When a contract is published, Java content classes are persisted in database tables. When a contract needs to be changed, tables are updated, and depending on the origin of update, XML representation is synchronized (via JAXB marshalling).

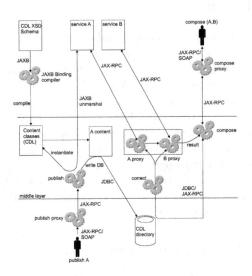

Fig. 5. JWSDP Runtime

JAX-RPC is used for communication with Web services. Since middle layer is also realized as a Web service, clients use JAX-RPC to invoke basic functions of the system, and middle layer uses JAX-RPC to invoke single or composite services. In JAX-RPC a remote procedure call is represented by an XML-based protocol, such as SOAP. Complex SOAP messages and their structure (envelope, encoding rules, conventions for RP calls and responses) are hidden by JAX-RPC API. This API supports development of server side (Web service implementation) and client side (Web service invocation) infrastructure. On the server side, remote procedures (Web methods) are specified by writing Java interface and one or more classes that implement that interface. On the client side, a proxy object is created that represents Web service. All Web methods are invoked on a proxy. Therefore, it is not necessary to generate or parse SOAP messages. The JAX-RPC runtime converts API calls and responses to and from SOAP messages.

The functioning of JAXB and JAX-RPC runtime is shown on Figure 5. It shows two typical use cases: publishing a new service to directory and composition of two services that are already in directory. Prior to any client calls, XSD schema describing Contract Definition Language is compiled with JAXB binding compiler, and content classes are stored in the middle layer. Client publishes new service by issuing SOAP or JAX-RPC call to the Publish Proxy, which delegates the call to the Publish service in the middle layer. A service that is to be published is located, and its CDL description is unmarshalled into precompiled content classes produced by JAXB compiler. Finally, write to underlying database is performed via JDBC which completes the publish process.

Composition is initiated by sending SOAP/JAX-RPC request to Compose Proxy, and the call is then delegated to Compose service in the middle layer via JAX-RPC. It processes composition request, retrieves partner service information from database using JDBC, verifies composition correctness by calculating function *correct*, and constructs required dynamic proxies that represent partner services using Dynamic Invocation Interface. Each proxy then connects to its implementation and middle layer coordinates message passing in a manner that depends on the composition pattern used. Result is returned to the client via Compose Proxy.

5.2   Service Directory and State Management

Service directory is realized as a relational database. There are several reasons why we use a relational database instead of a native XML database. Current XML databases still do not support W3C XML schema which we use to define CDL. Using native XML database could therefore lead to low data integrity. Furthermore, XML databases use XPath as query language, and it offers no support for grouping, sorting, cross document joins, and data types. Since service directory requires complex queries, this is a very limiting implementation factor. Still another downside is that updating requires retrieving an XML document, modifying it using own API and then returning it to database.

Database was designed to take full advantage of rich descriptive options offered by CDL in order to overcome UDDI limitations. The underlying database schema allows for searching for services directly, using any combination of properties defined in CDL. That means that it is possible to search for services by locations, methods they offer, classifications, and all other properties defined in their pre-conditions, post-conditions and invariants. One example query would be to find all services in the 1 km radius that accept postscript documents and print them in color with 1200 dpi resolution, free of charge if we can supply a security credential of certain type.

Up to now we have been talking about modeling Web services using abstract machines comprising state variables. It is obvious that we have implicitly assumed that some services can maintain their state between calls. However, Web services are stateless and we need to introduce state management mechanism.

Although Web services are inherently stateless, many of them allow for the manipulation of the state, such as persisting data into databases, file systems, or coordinating dependent messages. There is ongoing debate in the community whether Web services should or should not support state management. One view is that Web services are not another Object Request Broker architecture, and therefore should have no notion of state [18], while the other view is that state management plays the critical role in distributed computing and as such must be addressed at the architectural level [7]. Our position is that for the purpose of complex service interactions the latter view is correct. We identify two possible ways to associate a state with a Web service:

- A conversational service implements a series of operations where result of one operation depends on the prior operations of the same or other services. The state is maintained in the logical sequence of messages.
- A service that acts upon one or more persistent resources (database, file), creating, modifying or deleting it based on the messages it sends or receives.

Since conversational state can be implemented using WS-Coordination and WS-Context specifications, we concentrate on the interaction with stateful resources. Furthermore, we consider only relational database as a provider of background persistent resource. Interaction with persistent resource is described within the `resource` element of the CDL. Resource is identified by its name, uri, and resource manager (in our case, relational database driver). For each method acting upon a resource, one of the following actions can be defined: `CREATE`, `READ`, `MODIFY`, `DELETE`. Methods that create resources return resource identifier, while methods that read, modify and delete resources require resource identifier. Finally, resource property defines one or more CDL elements (state variables) that are bound to the underlying resource.

Our efforts in providing state management are compatible with the recent WS-Resource proposal [8], with the main difference being that WS-Resource supports broader range of persistent resources identified using WS-Addressing.

6   Conclusion

If Web services are to become the dominant architecture of future distributed systems, after connectivity is established (standardized) at least two more issues need to be supported at the architectural level: trustworthiness and automatic business to business (B2B) interactions. We try to address both in our proposed framework. We defined trustworthiness not only as security, but as an aggregation of properties (including but not limited to security) that composition process must guarantee. Therefore we adopted *correctness* as a term that best describes a "trustworthy" or a "trusted" composite Web service.

However, composition has still to be performed manually by application developer, albeit much easier and more flexible than in case of the other existing approaches as it now consists only of selecting appropriate services and applying composition operator (pattern) . The proposed framework offers possibility of true automatic B2B interactions [2]. Formal treatment of composition process enables use of various search strategies for the purpose of efficient allocation and verification in the process of automatic composition.

References

1. J.R. Abrial. The B Book. Cambridge University Press, 1996.
2. G. Alonso, F. Casati, H. Kuno, and V. Machiraju. Web Services: Concepts, Architectures and Applications. Springer-Verlag, 2004.
3. D. Berardi, D. Calvanese, D. G. Giuseppe, M. Lenzerini, and M. Mecella. Automatic composition of e-services that export their behavior. In Proc. of the 1st Int. Conf. on Service Oriented Computing (ICSOC 2003), 2003.
4. T. Bultan, X. Fu, R. Hull, and J. Su. Conversation Specification: A New Approach to Design and Analysis of E-Service Composition. In Proceedings of WWW2003, 2003.
5. F. Curbera, R. Khalaf, N. Mukhi, S. Tai, and S. Weerawarana. The Next Step in Web Services. Communications of the ACM, October 2003.
6. A. Ankolekar et al. DAML-S: Web Service Description for the Semantic Web. In Proceedings of the International Semantic Web Conference (ISWC), 2002.
7. I. Foster et al. Modeling stateful resources with web services. http://www.ibm.com/developerworks/library/ws-resource/ws-modelingresources.pdf, 2004.
8. K. Czajkowski et al. The WS-Resource Framework. http://www.globus.org/wsrf/specs/ws-wsrf.pdf, 2004.
9. X. Fu, T. Bultan, and J. Su. Formal Verification of E-Services and Workflows. In Proceedings of Workshop on "Web Services, e-Business, and the Semantic Web (WES): Foundations, Models, Architecture, Engineering and Applications", 2002.
10. R. Hamadi and B. Benatallah. A Petri Net-based model for Web Service Composition. In Proceedings of the Fourteenth Australasian database conference on Database technologies, 2003.
11. S. McIlraith and T.C. Son. Adapting Golog for Composition of Semantic Web Services. In Proceedings of the International Conference on the Principles of Knowledge Representation and Reasoning (KRR'02), 2002.
12. L.G. Meredith and S. Bjorg. Contracts and Types. Communications of the ACM, 46, No. 10, pp 41-47, October 2003.
13. B. Meyer. Applying Design by Contract. IEEE Computer vol. 25, no. 10, Oct. 1992.
14. Sun Microsystems. The Java Web Services Developer Pack. http://java.sun.com/webservices/downloads/webservicespack.html, 2004.
15. N. Milanovic and M. Malek. Current Solutions for Web Service Composition. IEEE Internet Computing, November/December 2004.
16. N. Milanovic and M. Malek. Extracting Functional and Non-functional Contracts From Java Classes and Enterprise Java Beans. In Proceedings of the Workshop on Architecting Dependable Systems (WADS 2004), Florence, Italy, 2004.
17. S. Narayanan and S. McIlraith. Simulation, Verification and Automated Composition of Web Services . In Proceedings of the International WWW02 Conference, 2002.
18. W. Vogels. Web Services are not Distributed Objects: Common Misconceptions about the Fundamentals of Web Service Technology. IEEE Internet Computing, November/December 2003.
19. J. Yang and M. P. Papazoglou. Web Component: A Substrate for Web Service Reuse and Composition. In Proceedings of 14th Conference on Advanced Information Systems Engineering (CAiSE02), Toronto, 2002.

Practical Approach to Specification and Conformance Testing of Distributed Network Applications*

Victor V. Kuliamin, Nickolay V. Pakoulin, and Alexander K. Petrenko

Institute for System Programming of Russian Academy of Sciences (ISPRAS),
B. Kommunisticheskaya, 25, Moscow, Russia
{kuliamin, npak, petrenko}@ispras.ru
http://www.ispras.ru/groups/rv/rv.html

Abstract. Standardization of infrastructure and services in distributed applications and frameworks requires ground methodological base. Design by Contract approach looks very promising as a candidate. It helps to obtain component-wise design, to separate concerns between developers accurately, and makes development of high quality complex systems a manageable process. Unfortunately, in its classic form it can hardly be applied to distributed network applications because of lack of adequate means to describe nondeterministic asynchronous events. We extend Design by Contract with capabilities to describe callbacks and asynchronous communication between components. The resulting method was used to specify distributed applications and to develop conformance test suites for them in automated manner. Specifications are developed in an extension of C language that makes them clear and useful for industrial developers and decreases greatly test construction effort. Practical results of numerous successful applications of the method are described. More information on the applications of the method can be found at the site of RedVerst group of ISP RAS [1].

Keywords: Design by Contract, asynchronous events specification, distributed system specification, formalization of standards, model based testing, conformance testing, automated test construction, specification extension of programming language, test oracle generation, UniTesK.

1 Introduction

Standardization of infrastructure and base services of distributed systems builds up its strength as the important component of the movement to availability and dependability of such systems. This process needs adequate support from methods and technologies of software construction. One of the promising approaches to development of high-quality complex software systems is Design by Contract (DbC) [2]. The key points of this approach can be stated as follows.

---

* This work is partially supported by RFBR grant 04-07-90386, by grant of Russian Science Support Foundation, and by Program 4 of Mathematics Branch of RAS.

M. Malek, E. Nette, and N. Suri (Eds.): ISAS 2005, LNCS 3694, pp. 68–83, 2005.

- Software is considered as a system of components separated from each other and communicating with each other only through the specified interfaces.
- An interface of the component is a set of its operations, which semantics is described with *preconditions* and *postconditions*. Precondition of an operation states the obligations of an environment – before the call of this operation a caller should ensure that the precondition holds. Postcondition states counter-obligations of the component. If the precondition holds just before the call of the operation, the component ensures that the postcondition holds just after the call. Preconditions and postconditions are usually formulated in terms of operation parameters and internal state of the component.
- Common parts of pre- and postconditions of all the component's operations can be stated as separate *invariants* representing integrity constraints on the component's state.

Design by Contract proposes a powerful and well-scalable software development method. It possesses the following advantages.

- Clear component boundaries and obligations make possible effective separation of concerns between different components, separation of development activities between their developers, and significant flexibility in their implementation.
- The approach ensures broad reuse. As long as we need some functionality stated as a postcondition, we can use any component providing this or more strict postcondition, if we in turn ensure the corresponding precondition. As long as developer can ensure some postcondition providing that the precondition holds, he or she may change the implementation of component without risk of introducing errors in the system.
- The approach applies rather uniformly to components of different scale. Subsystems consisting of many components can be also considered as components with their own contracts. With the help of contracts of a subsystem and constituent components we can ensure correctness of subsystem's decomposition, and so, step by step, can build rather complex systems on the same methodological base. The quality of the result can be predicted due to rigor of the approach combined with the simplicity of its application.

All this sounds great. Even more great it can be for modern service-oriented architectures, which are based on separate components providing services for each other. But Design by Contract in its classic form given in [2] can hardly be applied for modern complex software systems. We can formulate the following causes of this situation.

- Complex networking software uses many different kinds of communication activities between its components. For example, callbacks are rather common in distributed frameworks. Another widely used kind of communication between components of such systems is asynchronous events and messages. Consider these issues in more details.

Callback represents a parameter of functional type, constraints on which can be described only if we consider the properties of all functions that can be passed in this callback. So, we need to impose additional contract on callback, although it is only a parameter of some operation, not an operation itself. This kind of contracts and its use in system development is not concerned by the classic Design by Contract approach. Any time a callback parameter is used developers have to consider constraints on the corresponding operation outside of DbC framework or treat them rather informally.

DbC also has no special means to describe asynchronous communications, which is very important in modern software. Moreover, in DbC framework we can hardly find any means to reason about correctness of multiple asynchronous communications performed in parallel. This is really serious drawback of the approach, making it inapplicable to many modern systems.

– DbC approach was originally targeted for software design, and usually after coming to rather clear understanding of the system design designers and developers cannot get any more benefits from the contracts. So, the contracts, which require a lot of work to develop, become useless and are not supported after some phase of the project to minimize the total effort (sometimes they are also used for debugging). We think that to make contracts actually useful they need additional means to provide sound and full-scale quality control of the results of development performed on their base including automated test construction, test adequacy measurement, regression testing, and certification. The original approach says nothing about measurement of component's quality based on its contract – it provides only insights on possible usage of contracts to check runtime behavior of the components or to test them in a random fashion.

In this article we present possible solution of both problems. We provide an extension of DbC approach that adds just several new entities to original framework, but makes it applicable for specification of complex distributed applications and frameworks. In addition we present UniTesK test development technology, which used to construct conformance tests based on DbC specifications in automated manner.

In the next section the methodological base of the suggested approach is presented. Then we consider several practical applications of the extended Design by Contract to complex distributed systems, including both specification of system properties, formalization of the corresponding standards, and automated development of conformance test suites based on the stated specification. The fourth section presents a brief review of similar approaches to specification and test construction for distributed software. The last section of the article concludes the discussion and provides directions of possible future development.

2   Extending Design by Contract Approach

The main point of the presented approach is the same as of the original DbC – software is considered as a system of components communicating with each other through the specified interfaces. Interfaces consist of operations described by their pre- and postcondition. The differences begin when we deal with contract development for communication means of special kinds – callbacks and asynchronous events.

Callbacks. Callbacks are considered as parts of *inverse interface* – a kind of interface, which is used for calls from the system under consideration to its environment (cp. with usual direct interface used for calls from the environment to the system). So, a component implements some interface (its direct interface) and requires from the environment to support some inverse interface.

Operations in inverse interface are considered as ordinary operations and described by their pre- and postconditions. But when we define the behavior of an ordinary operation, which may make some calls to inverse interfaces (for example, it obtains callback as a parameter and its functionality requires to call this callback in certain situations), we should describe the constraints on these calls concerning their parameters and results.

To provide such a description we use *model trace* – each of components implementing inverse interfaces considered as storing a list of calls of its inverse operations. Each of those calls can be represented as a record with called operation identifier, values of its parameters, and value of the call result as fields. So, in postcondition of an operation using callback we can state that this callback was called with certain parameters. We also can state that the result of its call was used in a certain way to produce the result of the operation call.

This extension of DbC approach, although a minor one, provides powerful means to check systems interoperability or test whether the component can be used inside a framework. We should provide the system or the framework developed with description of contracts of both provided and *required* interfaces. To check that two systems can operate together we need to check that each one obeys the restrictions imposed by the other in preconditions of ordinary operations and postconditions of inverse operations. To check that a component can operate inside a framework we should test whether it ensures preconditions of operations it calls in the framework and postconditions of its own callbacks used by the framework.

Asynchronous events. More serious changes in usual DbC concepts are required to introduce asynchronous communications. Operations, whether they are performed synchronous or asynchronous calls, can be considered just in the same way. But asynchronous events are another kind of entities. We represent them as a special kind of operations without parameters, but having ordinary pre- and postcondition.

Precondition of an event describes situations when this event is valid. If the precondition does not hold, any occurrence of the event of this kind is incorrect. Postcondition of an event describes restrictions on data provided by the event. When precondition holds (so, events of this kind are possible) postcondition says whether this event provides correct data or not. Asynchronous messages can be also described in the same manner.

To define the correctness of a collection of events and calls occurring in the concurrent manner we use so called *interleaving* or *sequential semantics*. This semantics implies that the set of concurrent calls and events is performed in a correct manner if they can be performed in correct manner in some sequence. More precisely, a set $\{e_i, i \in [1..n]\}$ of calls of operations or occurrences of events performed on or provided by a component is considered to satisfy their contracts in a state s_1 of the component if there exists such a sequence $\{s_j, j \in [1..n+1]\}$ of component's states starting from s_1 and the corresponding ordering $\{i_j\}$ of those calls and events that each call or event e_{i_j} occurs in the state s_j, moves the component to the state s_{j+1}, and the contract of the corresponding operation or event holds for pre-state s_j, post-state s_{j+1}, provided values of operation parameters, and the result returned by the operation or by the event.

For example, if we have an operation printing "Hello, world!" on a printer and an event printing "Bye!", any result "Hello, world!Bye!" or "Bye!Hello, world!" is considered as correct result of concurrent call of the operation and occurrence of the event, but the result "Hello,Bye! world!" is invalid.

Although the proposed extension of DbC approach is not complex, it can be used successfully to describe distributed systems of practical significance, to obtain valuable results from more formal consideration of system properties, and to test the components of the system and a system as a whole, see the next section for examples of such applications.

Use of programming language extension. One more peculiarity of our approach is use of extensions of programming languages to specify software properties. This fact becomes important if one needs to apply some methodology or tool based on formal notation in industrial practice. Widely used programming languages are commonly recognized means of communication between developers and specifications written in their extensions are comprehensible for average software engineers. Specialized formal notations often require advanced mathematical education, do not contain adequate counterparts for widely used programming concepts (such as pointers), and therefore are rarely used in practice.

We propose uniform extension of C, Java, and C# languages [23] based on the main concepts of our approach – pre- and postconditions, invariants, asynchronous events, and inverse interfaces – and some additional syntactic sugar useful in postconditions, when one needs to work with both pre-states and post-states of the same objects. The main elements of the extension are as follows.

- Some operations in class (or some global functions in C) can have `specification` modifier saying that they contain contracts of the corresponding operations in the system under consideration. Such an operation can have *access constraints* describing the set of objects the operation has access to and the kind of this access (whether an object can be only read by the operation, only written, or both), *precondition* represented as additional block returning Boolean value, *postcondition* represented also as additional block also returning Boolean value. Postcondition has access to objects in the states preceding the call of the operation and the same objects in the states after the call (*pre-* and *post-states*). To refer a pre-value of a variable in a postcondition we can use *pre operator*.

In addition, specification operations may have `branch` constructs marking different behavior constraints and so defining specification-based coverage criteria for further testing.

- Operations marked with `reaction` modifier represent asynchronous reactions provided by the system. Such a reaction can also have access constraints, pre- and postcondition. But it has no functionality branches, since the behavior of the system in this situation is not determined by external input.
- Operations marked with `inverse` modifier represents inverse operations. They also can have access constraints, pre- and postcondition.
- Invariants are represented as special methods or functions marked with `invariant` keyword and returning Boolean result. The result says whether the invariant holds or not.

The code example 2 in Appendix demonstrates some elements of specification extension of C. It presents specification of a component implementing banking account. The component may be implemented as a web-service, or EJB, or plain class – this does not matter for the description of its functionality.

An account has two operations and can produce events on change of balance. Each event stores the difference between the new value of the balance and the old one. The first operation, `deposit()`, is used to deposit money on the account. The second one, `withdraw()`, is used to withdraw money from the account. Both operations can give rise to an event on account change storing the actually deposited sum or withdrawn sum as a negative number, but the results of several operations can be summed by one such event with the total change of the balance.

Negative value of the balance means that a credit is given to the account owner. The credit is limited by fixed maximum possible credit value, which is state in the invariant. Postconditions of operations and event define their impact on the state of the account. `bank` variable stores map of account identifiers into account structures.

3   Practical Applications of the Method

This section presents some results of practical application of the approach described above in two areas – clarification and formalization of standards and automated construction of conformance test suites for distributed software.

3.1   Formalizations of Standards

This subsection concerns with two case studies in standard formalization related with distributed applications. The first example is standard clarification and conformance test suite development for ISO/IEC 13818-11, a standard on Intellectual Property Management and Protection in MPEG-2 domain. The second one is a part of specification-based test suite development for an implementation of IPv6 protocol suite – the next generation of the Internet protocol.

Formalization of IPMP. A standard for MPEG-2 Intellectual Property Management and Protection (IPMP-2) [3] is an attempt to create a flexible and interoperable solution for Digital Rights Management in MPEG-2 distribution chain from content provider to user. For the sake of readability we will refer to ISO/IEC 13818-11 [3] as "IPMP-2 specification" below in this section.

The original architecture for protecting MPEG-2 movies, called Conditional Access (CA), proved to be non-interoperable. Playing content from a particular producers required purchasing Conditional access solution from certain vendors, and CA solutions from different vendors were incompatible.

IPMP-2 specification regulates IPMP operations on the side of a user. IPMP Device includes *a Terminal* and a number of *IPMP Tools.* IPMP Tools perform all operations needed to prepare data for playback such as user authorization, content deciphering, watermarks processing, etc. IPMP Tools are software or hardware modules that are plugged to specific *control points* in the MPEG-2 processing pipe. Terminal intercepts multimedia data and passes them to the corresponding instances of IPMP Tools for processing. Results of processing (e.g. deciphering) are returned to the Terminal for further processing. IPMP Tools interact with each other and the Terminal by means of message exchange. IPMP-2 specification provides a number of messages for several purposes, such as authentication or notification.

Content providers add *control information* and *protection signaling* to their content. This information includes indications on which tools to use, how to initialize the tools, etc. The IPMP Device parses content and tries to acquire IPMP tools from the network if needed. Then the device instantiates tools with given parameters and starts playback.

IPMP-2 specification uses Syntax Definition Language [4] for defining syntax of messages and IPMP-related data in content. Still the semantics of messages and data is defined in plain text without any formal notation.

The formalization of semantics of IPMP-2 operations has the following facets.

- Constraints on data integrity.
- Constraints on prerequisites and results of operations.

The work on IPMP-2 formalization was conducted for Audio Video coding Standard Working Group of China (AVS). Length of the studied specification is about 30 pages. The project resulted in two submissions [5,6] to AVS DRM group and a prototype of conformance test suite for processing IPMP Control Information in bit streams.

Other results of the project include the following.

- We identified significant inconsistencies in syntax specification of IPMP data in bit streams. For example, it allowed inserting up to 65 536 bytes of data (16-bit length field) in a descriptor which length is limited to 256 bytes.
- Under-specifications were found in the semantics of the Mutual Authentication – a security protocol for establishing trust between two tool instances. We demonstrated that current specification of Mutual Authentication does not ensure interoperability between implementations from different vendors.

- Correctness criteria of data in IPMP-2 specification are poorly defined. Discussion with IPMP developers showed that there are many implicit rules of what is correct and what is not. For example, IPMP-2 specification defines IPMP Tool List structure as a container for IPMP Control Info classes, but it is intended to carry information about tools only. We put this implicit constraint into explicit form: each element of IPMP Tool List is of IPMP Tool Info type. The list of constraints educed during the formalization for IPMP Control Information classes is presented in [6]. The constraints are not written in formal notation yet.

Taking into account numerous misspellings in code parts of IPMP-2 specifications the exact number of fixes we proposed is hard to count.

The standard study showed that IPMP-2 specification consists of several loosely related pieces that sometimes contradict to each other. Certain requirements are under-specified or contain errors.

Contract formalization of IPv6. IPv6 is a group of protocols located at the Network Layer of the OSI Reference Model [7]. IPv6 provides services to protocols of transport layer, such as UDP and TCP.

IPv6 features a much greater address space compared to IPv4, the current version of the Internet Protocol. Large address space enables true point-to-point connectivity within global scope. Besides extended address space IPv6 includes improved routing architecture and integrated suite of protocols for autoconfiguration and discovering the state of the communication.

Implementations of IPv6 provide three classes of interfaces: procedural (API), binary (ABI), and message-based.

Procedural interfaces include generic sockets API and several IPv6-specific extensions. Binary interfaces are non-standard, implementation-specific ways to access the kernel part of an implementation. Examples of such interfaces are request code for `ioctl` call on Unixens or control code for `DeviceIOControl` routine in Windows accompanied with memory layouts for inputs and outputs. Message-based interface is an abstraction for sending and receiving IPv6 datagrams to or from Data Link Layer.

IPv6 messages and part of procedural interface are standardized by Internet Engineering Task Force in IPv6-related *Requests for Comments* (RFCs). Binary interface and some part of procedural interface are not standardized and are implementation-specific. Since the component functionality should be understood unambiguously to apply Design by Contract fruitfully, it is natural to limit formalization to the scope of messages and standard API of IPv6.

The scope of our projects on IPv6 conformance testing was formalization and testing of the following basic features of IPv6.

- Sending datagrams from the transport layer to the network and processing of incoming IPv6 packets.
- Neighbor Discovery on hosts. Neighbor Discovery is a suite of service protocols for identifying router and neighbor nodes attached to a link and detecting their reachability status.

- Multicast Listener Discovery on hosts. Multicast Listener Discovery is a protocol to obtain information about multicast listeners attached to a link.
- UDP over IPv6.

The contract formalization is based upon requirements presented in regulating RFCs. We studied the requirements of many RFCs, most notably [8,9,10,11,12,13,14,15,16,17,18]. More than 400 separate functional requirements were elicited.

RFCs define protocol semantics in plain text mostly. Syntax is defined in tabular format with textual definition of bit-wise message layout.

We identified a number of inconsistencies and under-specifications in IPv6 regulating documents. For example, the specification of IPv6 protocol [8] enumerates a number of cases that should be considered as errors in incoming fragmented IPv6 packets, and a number of cases that are not errors. Unfortunately this enumeration misses several important cases, such as fragments overlap.

Despite the defects found we can state that IPv6 regulating requirements are well-defined as a rule. They are detailed enough to ensure interoperability between implementations and at the same time leave much flexibility to implementers.

The formal model of the IPv6 subset described above is about 8500 lines of code in the specification extension of C language [19]. The model was used to build a test suite that was applied to several open and commercial implementations of IPv6 protocol stack (see the next subsection).

3.2   Automated Conformance Test Construction

The historically first application of the extended DbC approach was automated test development. The specifications written in the described manner can be used to construct conformance test suite with the help of UniTesK technology. Here we provide a short introduction into UniTesK. The interested reader can find more details on it in [20,21,22,23,24].

The main principles of UniTesK test development may be summarized as follows.

- UniTesK is intended to develop conformance test suites automatically on the base of the specifications to be tested. The main approach to testing is black-box, testing adequacy is measured as the achieved during testing *coverage of specifications* according to some criterion. *Test oracles* – programs automatically checking the correctness of the behavior of the system under test – are generated automatically from contracts specified.
- User should manually write *test scenarios* providing very brief descriptions of the automaton model of the component under test, including structure of its state and the list of operations to be called in an arbitrary state. Each operation is supplemented with some procedure to generate values of its parameters. This procedure can be written manually or taken from a library; its main goal is to provide a large set of different arrays of operation

parameters values. The development of test scenario can be facilitated with the help of the template, taking several choices of the user as its input and generating all the other parts of the scenario. The main goal of a scenario is to ensure high level of test coverage in certain specification-based coverage metric.

Test scenarios provide a powerful feature – they can be used to process possible nondeterminism of specifications very effectively. To do this, one can define scenario states on the base of classes of states described in specifications. This technique called *factorization* allows creation of rather efficient and compact tests for complex subsystems. Details of the technique can be found in [25].

- Similar template technique is used to create test adapters providing binding between specifications and implementation under test.
- The UniTesK tool used translates specifications, adapters, and scenarios into the base language of the tool (C, Java, or C#) and executes the resulting test. During test execution the sequence of test calls is generated on-the-fly using the data presented in the scenario and the actual behavior of the system under test. The generation algorithm tries to call each operation in each state achieved, but do not perform calls that add nothing to already achieved test coverage in term of specifications (`branch` statements are an example of construct that can be used to define coverage of specifications).

UniTesK technology was used to develop conformance tests in the following projects.

- Development of regression test suite for switch operating system kernel for Nortel Networks. Results of this project was already presented in [20,24], see also [23]. Total size of the system under test is about 250 KLOC, the size of resulting suite of specifications and scenarios is about 140 KLOC. To our knowledge, this is the largest piece of formally specified software and the largest system tested in such a formal way. The total effort of the project is about 10 man-years, total duration – about one year and a half. 372 test scenarios were developed for about 500 procedures of the operating system kernel, 304 of those scenarios tested single procedure, 68 – a group of inter-operating procedures. With different parameters of execution the resulting test suite can perform from dozens of thousands to several millions of test cases. Several hundreds of defects were detected in critical telecommunication software already working in the field for about 10 years. Several of bugs found could cause cold restart of the system.
- Development of test suite and testing several IPv6 implementations. The detailed results those projects can be found in [19] and [22]. The projects also demonstrated the approach's capability to clarify ambiguous parts of informal telecommunication standards. The first project was conducted to test open IPv6 implementation of Microsoft Research. The results showed that the test suite provides good error detection – it found more errors that the counterparts we could compare with at that time (Microsoft Research organized an international contest in testing of this IPv6 implementation).

4 serious bugs were found in the system under test, one of them leads to operating system crash and can be used to shut down any remote node in IPv6 network. The second project is conducted in the Russian telecommunication software development company Octet by its own developers trained in our technology. It also resulted in several serious bugs found in another proprietary implementation of IPv6.

– Test development for a part of bank CRM system based on J2EE technology. This project demonstrated that UniTesK technology and tools can be applied to test distributed software constructed with the help of modern component-based technologies for multi-tier applications development. The duration of the project was about 2 months, and its results include about a dozen of bugs detected. The details of this and several other projects can be found on [23].

The diagram of process including standard formalization and conformance test suite development on the base of the approach presented is shown on Figure 1.

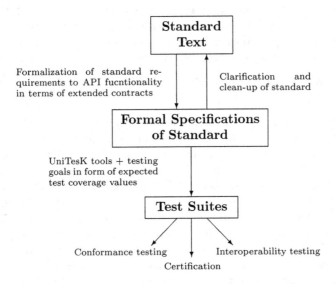

Fig. 1. Standard formalization and conformance test suite construction

4   Short Review of Similar Approaches

Here we present rather brief review of similar approaches taking into consideration only those that provide possibility to describe distributed systems formally and support test development automation for conformance testing, so a lot of interesting solutions stay out of scope of this section. More detailed and systematic review of various model-based testing techniques can be found in [26].

The most widely used practical approach to conformance test suite construction for distributed applications is based on informally determined *test purposes* and test cases manually developed on their base. In comparison with methods based on some formal description of application functions, it lacks strict and measurable definition of testing adequacy based on functional requirements and forces test developers to provide correct results only on the base of their understanding of the functions under test. Both disadvantages can be overcome by diligence and cross-checking, but not for large-scale systems.

The usual approach to formal specification and further testing of distributed software are based on some kind of transition systems – it may be labeled transition systems, input-output automata, and systems of communicating (extended) finite automata. Theoretical background for most part of those works was laid by J. Tretmans [27,28]. He proposed a formal definition of conformance relation between specifications and system under test and a method for test case generation based not only on possible inputs and outputs of the system under test, but also on special *quiescent* states where the system could not produce any output without some input from the environment. A series of tools based on those ideas were developed in the academic community, the most prominent from them are TGV [29] and TorX [30]. Some of those tools can take formal descriptions in such languages as SDL, LOTOS, or Estelle as input. In 2001-2003 years those tools were integrated into common environment developed in the AGEDIS project [31]. It includes uniform testing tool architecture and UML-based statecharts as standard input for such tools.

Transition systems used for automatic test generation proved to be very useful instrument, but they have the following disadvantages.

– State explosion problem. When one tries to model a real system on a detailed level, he obtains an unmanageable model with huge numbers of states and transitions. This is a demonstration of more serious drawback – transition systems can hardly be decomposed to separate different concerns and functions, they usually require considering the system as a whole to get valuable results. Design by Contract looks much more promising in this view since it provides a method to consider components of a complex system separately. In UniTesK state explosion problem can be overcome with the help of state factorization technique.

– Inefficient processing of nondeterminism. It is rather hard to introduce nondeterminism natural to distributed applications in transition systems and keep them useful. Most of them become inoperative after such a procedure. So, some special actions are always needed to introduce necessary nondeterminism in such a model. Contract based approach incorporates it naturally by stating the corresponding predicates in postcondition. Combination with factorization technique used in UniTesK, although not reducing concurrency-related nondeterminism to negligible level, makes it much more manageable.

5 Conclusion

The paper proposes an extension of Design by Contract approach for distributed network applications. The main extensions are constructs for specification of component-environment interaction through inverse interfaces and asynchronous events. Correctness of concurrent events is checked according to sequential semantics – a set of events is considered to be correct if and only if it can be ordered into a sequence conforming to all the contracts involved.

The extended DbC approach is used in practice-oriented UniTesK test development technology to construct conformance test suites in automated manner. UniTesK tools uses specifications in extension of programming languages (C, Java, and C# are supported now) to make them accessible and useful for ordinary industrial developers without background in formal methods. Although the approach and the test development technology based on it seems to be quite general, there are a lot of technical issues concerning their use in testing applications through GUI or Web interfaces, or through interfaces including timing events. Those issues should be resolved in future development. Since UniTesK tools were already successfully used in several industrial projects, the authors consider the proposed approach quite mature to be used in practical development of standards, distributed applications, and corresponding test suites.

The focus point of the approach presented is integrated process of standard formalization and conformance test suite development for it. This provides the following advantages.

- Standards are intended to state not only the common syntax of interfaces, but the common understanding of the functionality of the services described. Formalization removes a lot of ambiguities and misunderstandings, makes this functionality clearly stated, and so prevents a lot of potential problems with interoperability, sustainability, and dependability of future applications based on this standard.
- Formally stated functionality opens the door to automated conformance test suite construction, which decrease the effort to produce conforming applications and also make them more qualitative. In addition it gives a natural measure of testing adequacy in terms of requirements – one can precisely say now what is tested and what is not, to what degree some application conforms the standard and to what degree it breaks it.

Standards and development of infrastructure for distributed network applications attract more and more attention now. Neglect of the modern specification and automated conformance testing techniques has negative influence on both the quality of approved standards and the dependability of the systems developed on their base. Maybe, the same causes inhibit advancement of component-based development as a whole and growth of independent software vendors in particular. At the same time, the main restrictions of possible development are imposed not by the lack of adequate methods and tools, but by the lack of engineering staff having corresponding skills and experience in their application in practice. Our experience shows that this problem can be solved successfully.

Appendix

```
specification typedef struct account_model {
  int balance; int change; bool event;
} AccountModel = {};

invariant typedef AccountModel Account;

invariant int MaxCredit = 3;

invariant (MaxCredit) { return MaxCredit >= 0; }

invariant (Account * acc) { return acc->balance + acc->change >= -MaxCredit; }

typedef Integer AccountID;

Map * bank; // A map from Account ID to account

specification void deposit(AccountID *id, int sum) {
  Account * account = get_Map(bank, id);
  pre {
    return (sum > 0) && (account != NULL)
         && (account->balance + account->change < INT_MAX - sum)
         && (account->change < INT_MAX - sum);
  }
  post {
    return (account->balance == (account->balance))
         && (account->change == (account->change) + sum)
         && (account->event == true);
  }
}

specification void withdraw(AccountID *id, int sum) {
  Account * account = get_Map(bank, id);
  pre { return (sum > 0) && (account != NULL); }
  post {
    if(account->balance + account->change < sum - MaxCredit) {
      return (account->balance == (account->balance))
           && (account->change == (account->change));
    } else {
      return (account->balance == (account->balance))
           && (account->change == (account->change) - sum)
           && (account->event == true);
    }
  }
}

specification typedef struct account_notification {
  AccountID * id; int change;
} AccountNotification;

reaction AccountNotification * update() {
  Map * bank_saved = clone(bank); int i;
  pre {
    for (i = 0; i < size_Map(bank); i++) {
      if(((Account*)get_Map(bank, key_Map(bank, i)))->event) return true;
    }
    return false;
  }
  post {
    Account * account_saved = get_Map(bank_saved, update->id);
    Account * account = get_Map(bank, update->id);

    return (account_saved != NULL) && (account != NULL)
         && (update->change == account_saved->change)
         && (account->balance == account_saved->balance + account_saved->change)
         && (account->change == 0) && (account->event == false);
  }
}
```

Fig. 2. Example of specifications in C extension

References

1. http://www.ispras.ru/groups/rv/rv.html
2. Bertrand Meyer. Object-Oriented Software Construction, Second Edition. Prentice Hall, 1997.
3. ISO/IEC 13818-11:2004. Information technology – Generic coding of moving pictures and associated audio information – Part 11: IPMP on MPEG-2 systems. 2003.
4. ISO/IEC 14496-1:2001, Information technology – Coding of audio-visual objects – Part 1: Systems.
5. MPEG-2 IPMP Conformance Test Suite Development. AVS M1263: 2004/6.
6. Enhancing IPMP-2 for Conformance Testing. AVS M1487: 2004/12.
7. ISO/IEC 10731:1994. Information technology – Open Systems Interconnection – Basic Reference Model – Conventions for the definition of OSI services. 1994.
8. RFC 2460. S. Deering, R. Hinden. Internet Protocol, Version 6 (IPv6) Specification. December 1998.
9. RFC 2461. T. Narten, E. Nordmark, W. Simpson. Neighbor Discovery for IP Version 6 (IPv6). December 1998.
10. RFC 2462. S. Thomson, T. Narten. IPv6 Stateless Address Autoconfiguration. December 1998.
11. RFC 2463. A. Conta, S. Deering. Internet Control Message Protocol (ICMPv6) for the Internet Protocol Version 6 (IPv6) Specification. December 1998.
12. RFC 2464. M. Crawford. Transmission of IPv6 Packets over Ethernet Networks. December 1998.
13. RFC 3513. R. Hinden, S. Deering. Internet Protocol Version 6 (IPv6) Addressing Architecture. April 2003.
14. RFC 2373. R. Hinden, S. Deering. IP Version 6 Addressing Architecture. July 1998.
15. RFC 2292. W. Stevens, M. Thomas. Advanced Sockets API for IPv6. February 1998.
16. RFC 2553. R. Gilligan, S. Thomson, J. Bound, W. Stevens. Basic Socket Interface Extensions for IPv6. March 1999.
17. RFC 2675. D. Borman, S. Deering, R. Hinden. IPv6 Jumbograms. August 1999.
18. RFC 2710. S. Deering, W. Fenner, B. Haberman. Multicast Listener Discovery (MLD) for IPv6. October 1999.
19. http://www.unitesk.com/products/ctesk/
20. V. Kuliamin, A. Petrenko, I. Bourdonov, and A. Kossatchev. UniTesK Test Suite Architecture. Proc. of FME 2002, LNCS 2391, pp. 77–88, Springer-Verlag, 2002.
21. V. Kuliamin, A. Petrenko, A. Kossatchev, and I. Bourdonov. UniTesK: Model Based Testing in Industrial Practice. In proceedings of 1-st Europpean Conference on Model-Driven Software Engineering, December 2003.
22. V. Kuliamin, A. Petrenko, N. Pakoulin, I. Bourdonov, and A. Kossatchev. Integration of Functional and Timed Testing of Real-time and Concurrent Systems. Proc. of PSI 2003, LNCS 2890, pp. 450–461, Springer-Verlag, 2003.
23. http://www.unitesk.com
24. I. Bourdonov, A. Kossatchev, A. Petrenko, and D. Galter. KVEST: Automated Generation of Test Suites from Formal Specifications. FM'99: Formal Methods. LNCS 1708, Springer-Verlag, 1999, pp. 608–621.
25. I. B. Burdonov, A. S. Kossatchev, and V. V. Kulyamin. Application of finite automatons for program testing. Programming and Computer Software, 26(2):61–73, 2000.

26. V. Kuliamin. Multi-paradigm Models as Source for Automated Test Construction. Proc. of Workshop on Model Based Testing, Barcelona, Spain, March 2004. Also available in Electronic Notes in Theoretical Computer Science 111:137–160, 2005, Elseveir.

27. J. Tretmans. A Formal Approach to Conformance Testing. Proceedings of the IFIP TC6/WG6.1 Sixth International Workshop on Protocol Test systems, Pau, France, September 1993, pp. 257–276.

28. J. Tretmans. Test Generation with Inputs, Outputs and Repetitive Quiescence. Software – Concepts and Tools, 17(3):103–120, 1996.

29. J. -C. Fernandez, C. Jard, T. Jéron, and C. Viho. Using on the fly verification techniques for the generation of test suites. Proccedings of CAV'96, Conference on Computer Aided Verification, Rutgers University, New Brunswick, New Jersey, USA, July-August 1996.

30. J. Tretmans, A. Belinfante. Automatic testing with formal methods. In EuroSTAR'99: 7-th European Int. Conference on Software Testing, Analysis and Review, Barcelona, Spain, November 8-12, 1999. EuroStar Conferences, Galway, Ireland. Also: Technical Report TRCTIT-17, Centre for Telematics and Information Technology, University of Twente, The Netherlands.

31. http://www.agedis.de/

Model-Based Optimization of Enterprise Application and Service Deployment

András Balogh, Dániel Varró, and András Pataricza

Budapest University of Technology and Economics,
Department of Measurement and Information Systems,
H-1117 Budapest, Magyar Tudósok körútja 2
{abalogh, varro, pataric}@mit.bme.hu
http://www.inf.mit.bme.hu/FTSRG

Abstract. Enterprise services play an important role in these days' business environments. With the growing incidence of web services, the web service-based collaboration of systems is spreading. This leads to a large number of depending services. As these components form critical business applications, the availability and performance aspects of them are critical. We introduce in this paper a method that collects the QoS requirements of the high level services and propagates them through the dependencies to lower levels. Our tools also generate an optimal deployment configuration to a definite set of server nodes that guarantees the required availability and performance characteristics for all services.

1 Introduction

Nowadays, enterprises heavily depend on the quality of the service (QoS) they provide. In many cases, this quality of service primarily depends on the quality of their business IT systems. Such applications not only have to deliver a service with correct functionality (e.g. a bank transaction withdraws the right amount of money from our bank account), but these services has to meet several non-functional requirements (e.g. bank customers would expect the system to be available when they access it).

While non-functional requirements (such as performance, reliability, and availability) play an important role in business IT systems, QoS issues are neglected when designing such systems. Typically, the QoS assessment of a system is deferred until the deployment phase, which is frequently too late: if the deployed system does not meet its QoS requirements, it will cause an immense increase in the cost of the project.

To avoid these risks, the fulfillment of these QoS attributes has to be validated throughout the entire lifecycle of the project. Due to the increasing success of the Model Driven Architecture (MDA) [1], such a validation preferably starts from a model-based estimation / prediction of the QoS parameters carried out in a very early phase of the design.

Enterprise systems consist of many heterogeneous hardware and software components that form a logically and physically distributed infrastructure. The prediction

M. Malek, E. Nette, and N. Suri (Eds.): ISAS 2005, LNCS 3694, pp. 84–98, 2005.

and calculation of the QoS attributes in such an environment is difficult, because of the high number of dependencies between software components.

In the current paper, we present a method for model-level calculation of availability and capacity requirements for enterprise services and application components. Our method takes the QoS attributes of the highest level services that are directly accessed by users and propagates these values to the lower levels. Using a hardware catalog, we also synthesize the optimal hardware architecture for running the services while maintaining the required availability and performance characteristics.

We illustrate our results with an example on Java 2 Enterprise Edition (J2EE) platform that is widely used and supported by software vendors like IBM, SUN, BEA, and many more.

2 Standards and Technologies

The development methodology of enterprise systems usually integrates several technologies and standards in the fields of design and implementation. We introduce the most commonly used of these in the following sections.

2.1 Model Driven Architecture

MDA[1] (Model-Driven Architecture) is an emerging concept of the OMG (Object Management Group). Its main goal is to provide a framework for model-based system development, even in rapidly changing hardware and software environments. In particular, MDA addresses the challenges of todays highly networked, constantly changing system environments by providing an architecture that assures cross-platform interoperability, portability and reusability of software components.

MDA recommends starting the design with a platform-independent model (PIM) of the application. This focuses on the functional requirements, business logic, and the logical data structures, independent from any implementation technology e.g. J2EE. The suggested modeling language is UML 2 [2].

The next (automated) step is transforming the model to one or more PSMs (Platform Specific Model), which now contains information about the running middleware and other platform components.

The final step of the MDA design flow is the code generation phase. This is a largely automated process, which yields the source code of the application. The developers can extend the generated code with manually written parts.

The main advantage of MDA is the portability of software components between platforms, without manually recoding the application. This reduces the costs and time-to-market of the new versions, while reducing the probability of errors and potential security leaks caused by manual coding.

2.2 Enterprise Services

In addition to delivery the proper functionality, enterprises today need to extend their reach, reduce their costs, and lower the response times of their services to customers, employees, and suppliers.

Typically, applications that provide these services must be integrated with the existing enterprise information systems (EISs) with new business functions that deliver services to a broad range of partners. The services should be highly available, to meet the needs of today's global business environment; secure, to protect the privacy of the business partners and the integrity of the enterprise; reliable and scalable, to ensure that business transactions are accurately and promptly processed, and business growth can be followed by the software and hardware infrastructure.

The second important aspect of services besides the functional requirements is the quality-of-service (QoS) attributes of the services. Services are commercial in most cases, so the availability and proper performance of the services is an important point. The guaranteed QoS attributes are defined in a Service Level Agreement (SLA). This acts as a contract between the service provider and the user.

In most cases, enterprise systems are implemented as distributed multi-tier applications. The middle tier functions are grouped into web services and can be automatically discovered and used by partners, allowing the automatic intra-enterprise collaboration. This leads to a distributed, multi-organization service oriented architecture (SOA) [3] that involves many partners and service endpoints.

Several standards have been developed to ease the integration of basic services into complex processes. One of the most commonly used ones is the Business Process Execution Language (BPEL) [4] that is supported by the greatest software and middle tier vendors. BPEL can be used with services running on various platforms, such as Microsoft .NET and J2EE.

2.3 Design for High Availability in J2EE

We introduce the basic architecture and common redundancy patterns of the Java 2 Enterprise Edition (J2EE) platform. This introduction is based on the 1.4 version of the J2EE specification [5].

2.3.1 J2EE Architecture

The basic architecture of J2EE is built up from at least three tiers.

The first tier is responsible for data persistence. This layer consists of so called *entity beans*. Beans represent the smallest independent software components in Java. An entity bean maps to a row in a relational database table. The entity beans and their EJB container manage the creation, storage, and retrieval of application data. The architecture also defines transaction support for bean methods.

The second tier of the architecture is responsible for the implementation of business logic. This tier is made up from *session beans* that collect the business methods required by the application logic and may also contain *message-driven beans* that support asynchronous communication with reliable messages.

The third (presentation) layer of the architecture is responsible for the implementation of application user interfaces. Web clients use the HTTP interface of the server to access the web pages that are dynamically created by the servlet container of the J2EE server.

The J2EE architecture defines the infrastructural services that are needed to execute enterprise applications therefore the developers do not need to create custom interfaces to naming, authentication, message queuing, and database servers.

2.3.2 Redundancy Patterns

There are several patterns for creating redundant J2EE server architecture for high availability and load balancing solutions. The basic concept behind these techniques is clustering. A cluster is a set of computers which all run the same J2EE application and communicate with each other to determine the set of currently active nodes and to synchronize their internal state.

The incoming requests are distributed between the running nodes to provide load balancing. If a node goes down, the other nodes take over its workload. This results in higher availability, because the failure of a single node does not directly affect the availability of the services and applications.

2.3.3 Existing Development Environments

Today's development environments (such as IBM Websphere Studio and Microsoft Visual Studio .NET) focus on modeling the functionality of applications, and the generation of the source code skeletons for software components. They do not support, however, the definition and evaluation of QoS attributes such as availability and performance. The evaluation of the non-functional parameters is deferred to the testing phase of the development.

The capacity design of the hardware infrastructure environment that runs the application is not supported by any automated tools; therefore designers have to manually create the deployment plan and tune the hardware infrastructure to achieve the needed availability and performance levels.

2.4 Fault Model

Enterprise application server nodes consist of several layers of hardware and software components. We assume that errors can only occur in lower levels (illustrated by Figure 1), either in the hardware or in the operating system level. Errors of the higher level components can be easily and rapidly detected and repaired by the local management agent that can restart the failed component.

Hardware and operating system errors cannot be repaired as fast as the higher level errors. This results in a much longer downtime. Even if the severity of the hardware errors is lower than the software errors, the overall service downtime is much higher because of the longer repair time.

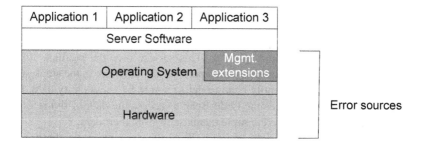

Fig. 1. Application server components

Our fault model is applicable if we can suppose that the higher level software components are stable enough not to cause a significant downtime. Commercial J2EE application server software and database management software meet this requirement. As the application modules are typically generated by automatic code generators from a higher level model, such as BPEL (Business Process Execution Language), we suppose that the application components cause no errors.

The fault model introduced here has several limitations. If running on a dependable, highly redundant hardware, where hardware component and operating system errors do not cause system restart (for example, in a massively parallel system) higher level software errors will be dominant in service downtime.

As in most cases business software components are running on entry or medium level servers where the fault hypothesis is satisfied.

3 Modeling Technique

We introduce the modeling techniques used for representing the functional and QoS aspects of the systems in the following sections. We used the standard UML Profile for J2EE for functional modeling with several extensions to allow the representation of the QoS aspects of system components.

We illustrate the explained concepts with a running example.

3.1 Running Example

Our example is a simple order processing and stock management system. It receives orders from customers, prints invoices and generates backorders to part suppliers if necessary. It consists of several services.

The *partner* service is responsible for storing and retrieving the partner data, such as name, address, payment and discount options. The *product* service offers access to the various data of the companies' products such as name, price, and description.

The *stock* service manages the administration of the product's stocking and movements. It relies on the product service. The actual stock state can be queried for a specific product, and goods movements can be administered. The *accounting* service is used to create invoices for customers who order goods from the company. This service relies on the partner service. The *ordering* service that relies on the product, the stock and the accounting services manages the incoming product orders.

The *backorder* service is responsible for the creation of backorders to parts suppliers if a specific product runs out of stock. This service uses the product, stock and partner services. It is automatically invoked periodically and checks the stock state.

Each service bean uses an entity bean to get access to business entity data. For example, the ordering service uses the OrderBean to access the data of living orders in the system. All entity beans use the same database to store their data.

The services are grouped into three EJB containers and a database module. Figure 2 illustrates the deployment units of the system.

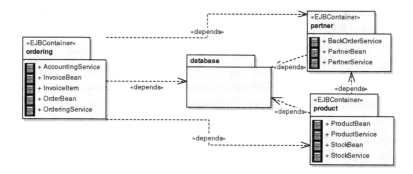

Fig. 2. Deployment units in the system

3.2   Modeling J2EE Components

Modeling J2EE components in UML is standardized by several UML Profiles, for example the UML Profile for EJB [6]. The standard UML classes are extended with stereotypes and tagged values to provide information about the J2EE-specific properties of the system components.

The EJB profile defines several stereotypes for marking the various types of Enterprise Java Beans. The "SessionBean" stereotypes marks the session beans, and the "EntityBean" marks the entity beans. Several other types (for example, message driven beans) and subtypes (container, or bean managed persistence) of components can be defined, but these are only necessary for the code generation, not for the QoS analysis.

3.3   Modeling Non-functional Requirements

3.3.1   Representing QoS Attributes in UML Models

Non-functional requirements are out of the scope of the EJB profile described earlier; therefore these properties have to be modeled in another way. Modeling non-functional aspects of systems is described in UML Profile for Schedulability, Performance, and Time [7], and in UML Profile for Quality of Service and Fault Tolerance [8]. These profiles define elements for the specification of non-functional (for example, performance and availability) parameters of system components.

In our architecture, the service access points are either web services (represented by stateless session beans) or session beans (either stateful or stateless). This means that the QoS attributes are defined for these components, and has to be automatically propagated to lower level ones. Other session beans and entity beans work at lower levels to provide basic services for the others and provide access to databases.

- The QoS attributes that are used in our work are the expected availability and the peak workload of services.

These two attributes are specified as tagged values (*QOS_Availability* and *QOS_Workload*, respectively) for the components. Our optimization method also needs the component dependencies to be defined, with the help of standard UML dependencies. This way, our transformation can compute the needed QOS aspects of the lower level components.

As the unit of deployment in J2EE is the EJB module, which is a set of Enterprise Java Beans, we need to propagate the QoS attributes of beans to these modules. EJB modules are represented by UML packages in our models. A package gets the maximal availability requirement and the sum of the peak workload of its components. These values are used in the further calculations.

3.3.2 QoS Attributes in the Example

Not all services in our example have QoS attributes, because the source model contains only those attributes that are defined for the external available, complex services. The attributes for the other services will be automatically calculated by the optimizer.

As mentioned before, the QoS attributes are propagated to the EJB modules. The results are illustrated in Table 1 (N/A means that no explicit constraints are defined).

Table 1. Calculated QoS parameters for the EJB modules

Module name	Availability	Workload
Product	99.9	200
Partner	99.9	22
Ordering	99.99	50
Database	n/a	n/a

3.4 Modeling Available Physical Components

In our scenario physical system components are server computers that can run J2EE applications. UML Components represent the server types in our model.

3.4.1 Performance Metrics

There are several industrial standard benchmarks that measure the overall performance of a server system with all of its hardware and software components. One of these is the TPC-W benchmark developed by the Transaction Processing Performance Council. This test measures the performance of a web-based transactional system. As enterprise services are web services, this benchmark can be used as a reference for the overall system performance.

The model of the server components has a tagged value called *"performance"*, which holds the number of the served requests determined by the TPC-W benchmark. This will be used to determine the capacity (maximum workload) of the server.

3.4.2 Component Costs

The server components also have an associated cost value that indicates the TCO (Total Cost of Ownership) value of the server, including the cost of all hardware (processor, memory, disks, UPS, and so on) and software (OS, application server, management tools) components, and all associated services (extended warranty, on-site service) for a given period of time.

The time factor depends on the desired lifetime of the service or application that is served. In case of applications with long life cycle, the basis of the calculation could be the expended life cycle of the server farm. The typical length of the lifecycle of servers is around 2-3 years. To make the cost of the possible server choices comparable, the time factor should be universal for the whole model.

The components have a tagged value called "*TCO*" to hold the Total Cost of Ownership value.

3.4.3 Component Availability

The third QoS value that is attached to servers is the availability. This attribute depends on the hardware, software, and also on the value added services offered to the specific server. Hardware suppliers specify the MTBF (Mean Time Between Failures) value for computer hardware. This can act as a starting point of availability calculation. As mentioned before, we handle only hardware and operating system errors, as the potential downtime they can cause is much higher than is case of higher level software component errors (application server or database server components), because the software components can be efficiently monitored and restarted in case of errors.

If we want to achieve high availability, software errors play also an important role, as the 30-60 sec typical restarting time of a J2EE application server can also affect the availability of a critical service. In this case, the system adds extra redundancy to avoid the unavailability of service.

Availability (A) can be calculated from the MTBF value and the MTR (Mean Time to Repair) by the following formula (1).

$$A = MTBF/(MTBF/MTR) . \qquad (1)$$

Availability is also attached to the server components by a corresponding tagged value.

3.4.4 Component Cardinality

The last attribute of physical system components that is required for our analysis is the maximum number of available instances of a given server type. This is important if we want to deploy the needed services on an existing infrastructure. The number of the server instances is stored in tagged value "*max_instances*".

3.4.5 Physical Components in the Sample System

In this sample system we have three different server machines that can be used for serving the application. The performance and availability data of the servers are approximate values as we do not know the exact service contracts data and resale prices for these machines.

The first configuration is an entry level Intel x32 server that can process 90 requests per minutes and has an availability of 97%. Its TCO is 2500 Euros.

The second configuration is a more robust Intel x32 server that can process 170 requests per minutes and has an availability of 98%. Its TCO is 3700 Euros.

The third configuration is a robust multi processor PowerPC server with redundant components and can process 1400 requests per minutes and has an availability of 99.9%. Its TCO is 18000 Euros.

4 The Optimization Workflow

We introduce our code generation and architecture synthesis methods in this section. We have been used our general-purpose model transformation system for the implementation of the required transformations and code generation scripts.

4.1 Architecture Synthesis

The process synthesis consists of two main steps (see Fig. 3). The first step is the transformation of the UML model of the system to a special format that can be imported to the optimization program. The program that is used for the synthesis is the second step of our workflow. The result of the process is the recommended architecture of the system. In parallel with the optimization, the source code of the system components can also be generated with commercial code generators or the VIATRA 2 framework.

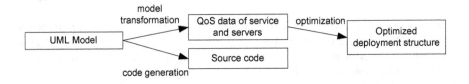

Fig . 3. Model processing workflow

4.1.1 Model Transformation
The first basic step of the transformation is the propagation of the bean QoS values to the EJB modules as described earlier. Each EJB module inherits the maximum availability and aggregates the performance value of its beans.

The second step is to propagate the bean dependencies to the modules. An EJB module depends on an other one if at least one of its beans depends on one of the beans in the other container.

After all QoS attributes and dependencies have been propagated to EJB containers, the transformation program generates the input file for the optimization program by the traversal of the UML package structure. It prints out the defined capacity and availability requirements and dependencies for every UML package that is marked with the stereotype EJBModule.

4.1.2 Deployment Optimization
We developed a simple command-line application that computes the optimal deployment pattern for the input system. It takes the input file with the system services and available hardware components and generates the optimal system configuration as output.

The arrangement of EJB modules between servers is a special optimization task. There are finite number of resources and finite number of software modules that must be related to each other. There are also several special constraints that describe the QoS constraints and dependencies of the components.

The goal of the optimization process is to minimize the overall system cost while providing the necessary system availability and capacity.

4.2 Implementation Technology

We have developed VIATRA [9], our general model transformation system, in order to support the dynamic, multilevel metamodeling features of VPM [10], and generic/meta transformations [11]. The main intended usage of our framework is dependability evaluation and optimization of business process workflow models and UML models.

A source user model (which is a structured textual representation such as an XMI description of a UML model exported from a CASE tool) is imported into the VPM modelspace. Transformation specifications can be constructed by combining graph transformation [12] and abstract state machine [13] rules. These rules can be created within the framework or in a UML tool using a special profile (and, in the future, using the QVT standard).

The rules are then executed on the source VPM model by the generic (higher-order) VIATRA rule interpreter in order to yield the target (VPM) model. Finally, the target model can be serialized into an appropriate textual representation specific to back-end tools.

The VIATRA 2.0 framework is implemented as a set of plugins for the Eclipse framework [14] that is a widely used open-source system development and modeling framework.

5   The Mathematical Model for Optimization

Optimization, in general, means a method that searches a point in the problem space that satisfies the defined *constraints*, and the *objective function* has a maximum (or minimum) value. Several special optimization problem classes have been defined, for example the traveling agent problem.

5.1 Our Optimization Problem

5.1.1 Initial Steps

The first step of the optimization process is the calculation of the aggregate workload of software modules. The developer only has to specify the direct workload for a specific container (the actual requests from clients) but the capacity needs to depend also on the indirect workload (calls from depending services). In our simple model, we suppose that a dependency represents a single call to the target service.

The calculation of aggregate workload is a recursive expression that calculates the workload as a sum of the direct workload and the additional workload of depending services (expression (1)). The *depends(i)* is a set of services that depend on service i.

$$\text{Workload}(i) = \text{Capacity\_need}(i) + \sum_{j \in \text{depends}(i)} \text{Workload}(j) \cdot \tag{1}$$

5.1.2 The Workload Constraint

The workload constraint means that the aggregated capacity of all deployed software modules on a specific machine must not exceed the capacity of the machine. A further tuning possibility is to define a saturation factor (SF) that specifies the maximum rate of workload on machines. Expression (2) specifies the workload constraint.

$$\forall m \in HW : Capacity(m) * SF \geq \sum_{s \in deployed(m)} Workload(s) \tag{2}$$

5.1.3 The Availability Constraint

The availability constraint specifies that the actual availability of each service must be at least as high as the required availability. The actual availability of a service can be calculated from the availability of the hardware that runs the service and the availability of depending services. Expression (3) specifies the availability constraint.

A service is available if the hardware it is running on is available and all the required services of the specific service are available. We suppose that if a hardware unit is running then all services deployed on it are running as well. We also suppose that all hardware nodes are independent, which means that all of them have their own uninterruptible power supply, disk subsystem, and so on.

$$\begin{aligned} A_{act}(i) &= P(\text{HW available} \wedge \text{All needed services available}) = \\ &= P(\text{HW available}) * \prod_{\text{all HW running required service}} P(\text{HW available}) = \\ &= A_{HW(i)} * \prod_{\forall j, HW(j) \text{ running needed service}} A_{HW(j)} \quad \forall i \in \text{services} : A_{act}(i) \geq A_{required}(i) \end{aligned} \tag{3}$$

5.1.4 The Objective Function

The *objective function* of the optimization process is the overall cost of the system, as described by expression (4). The total cost of the system is the aggregation of the product of the cost and the actual number of the defined hardware components.

$$TCO_{System} = \sum_{m \in HW} TCO(m) * number\_used(m) \cdot \tag{4}$$

5.1.5 The Solutions

A solution of the problem is a mapping between computers and software modules that satisfies all constraints. Solutions are computed by a backtrack algorithm that tries to build the mapping step-by-step while maintaining the constraints. The optimal solution is the solution that has the lowest overall cost.

5.1.6 Additional Steps

If the required availability or performance levels cannot be reached using the basic hardware types defined in the model, the optimizer applies the J2EE redundancy patterns for the design. This means that the program creates clusters from the basic hardware nodes to raise the availability and performance of a server. If the availability requirements do not allow single point of failures in the system, the developer can specify that only redundant arrays of machines can be used during architecture synthesis. This means that the program creates clusters even if the performance and availability of a single computer could satisfy the needs of the services.

The capacity of a cluster consisting of several nodes can be lower than the sum of the capacity of the nodes. That is because various synchronization messages and algorithms that are running on nodes. The typical value of performance loss depends highly on the actual server software, but it can be measured or taken from server benchmarks. Our tool supports the definition of a "performance loss percent" that is subtracted from the sum performance of the cluster nodes. If the services only use stateless session beans and entity beans, this loss is negligible in most cases.

More components (EJB containers) can be deployed on the same server if the hardware has enough capacity for running all the services. This ensures that the workload of the servers will be nearly equal, and the hardware costs will be minimized.

The optimization program calculates the optimal configuration of services and hardware nodes using the explained equations and constraints. The output of the program is a list of services and the associated hardware nodes. This defines the suggested configuration of the system.

5.2 Optimization Results of the Example

The optimal configuration with the original QoS attributes is to create a four node cluster from the medium level server. This configuration has an availability of more than 99.99999% and a total cost of 14800 Euros. All services are deployed to this single cluster.

If we suppose that the business grows very rapidly and the workload grows to the tens of the original. The optimal architecture in this case is to create a two node cluster form the third server that runs the database, and the partner modules, an other two node cluster from the third type that runs the product module, and a three node cluster formed from mid range servers that runs the ordering module.

The total cost of the system is 83100 Euros. This is 5.6 times more than the original, but offers 10 times more performance. This shows that a few of large but expensive servers can be used for serving heavy workloads, but for small workloads clusters built up from cheap servers can be used successfully.

6 Related Work

The model-driven analysis of QoS attributes of component-based systems under design has recently become a hot research topic. Primary focus is usually put on performance issues such as, e.g., in [17,18,19]. The early assessment of traditional dependability attributes is carried out in [20,21]. In most of these papers, a traditional transformation-based approach is followed where the QoS parameters are generated from a higher-level initial model (semi-)automatically. In contrast to these approaches, we focused on availability and cost parameters of deployment.

In [22], the authors define a method for dependability analysis of systems based on UML models. The basic idea behind that method is the transforming UML models to Timed Petri Nets (TPN). The starting point of the method is the architectural level model, so it works on a static infrastructure and does not modify the systems architecture.

In [23], the authors define a method for dependability analysis of systems based on UML models. The basic idea behind that method is the transforming UML models to Timed Petri Nets (TPN). The starting point of the method is the architectural level model, so it works on a static infrastructure and does not modify the systems architecture.

Probably, the most closely related work is that work of Bastaricca et al. [15], where the authors describe two deployment optimization methods that can be used in a distributed component-based environment. Both algorithms do the optimization of the deployment, but they work on a static infrastructure that cannot be modified. This way, they cannot be used for infrastructure planning, only for deployment on existing hardware environments. Moreover, the algorithms do not optimize for TCO, but for network utilization.

7 Conclusion and Future Work

Enterprise services play an important role in today's business environment. Besides the functional requirements the quality-of-service attributes are also more and more important. The most commonly used development environments do not support the handling of QoS attributes like availability and performance requirements of services.

In the paper, we introduced an approach to generate the optimal deployment plan for a set of enterprise services based on the UML model of the system and a hardware specification catalog. Our method ensures that the deployed system will keeps the availability and capacity constraints defined by the system model.

The current method is applicable only in design time, thus further improvements has to be made to extend its capabilities to allow the runtime reconfiguration of the systems. This will enable the automatic tuning of system availability and performance reflecting to the changes in the environment (the growth of the workload or the permanent fault of a server node).

To achieve this functionality, our optimizer need to be connected to a systems management software such as IBM Tivoli [16] that collects runtime information about the usage statistics and state of services and hardware nodes.

Further research has to be done for discovering methods to a finer granularity workload prediction that relies on the behavioral model of the services (for example it discovers that a service uses another several times). Other methods has to be developed to predict the relative weights of service executions to distinguish more complex services as they cause higher workload as simple services.

References

1. The Object Management Group, *MDA Information Portal*, http://www.omg.org/mda
2. The Object Management Group, *UML2 Superstructure specification*, August 2003 http://www.omg.org/
3. Steve Graham et al, *Building Web Services with Java: Making sense of XML, SOAP, WSDL and UDDI*, 2002.
4. Microsoft, IBM, BEA, et al. *Business Process Execution Language for Web Services Specification*. 5 May 2004.
5. Sun Microsystem. *Java 2 Platform Enterprise Edition Specification v1.4*. November 2003. http://java.sun.com/j2ee
6. Jim Conallen, Building *Web Applications with UML*, Addison-Wesley, 1999.
7. The Object Management Group, *UML Profile for Schedulability, Performance, and Time Specification*, January 2005.
8. The Object Management Group, *UML Profile for Modeling Quality of Service and Fault Tolerance Characteristics and Metrics*, September 2004.
9. D. Varró, G. Varró, and A. Pataricza. *Designing the automatic transformation of visual languages*. Science of Computer Programming, vol. 44(2):pp. 205-227, 2002.
10. D. Varró, A. Pataricza. *VPM: A visual, precise and multilevel metamodeling framework for describing mathematical domains and UML*, Journal of Software and Systems Modeling vol. 2 pp. 187-210, 2003.
11. D. Varró, A. Pataricza. *Generic and Meta-Transformations for Model Transformation Engeering*. In Proc. UML 2004: 7th international Conference on the Unified Modeling Language
12. H. Ehrig, G. Engels, H.-J. Kreowski, and G. Rozenberg (eds.). *Handbook on Graph Grammars and Computing by Graph Transformation*, vol. 2: Applications, Languages and Tools. World Scientific, 1999.
13. E. Börger and R. Stark. *Abstract State Machines. A method for High-Level System Design and Analysis*. Springer, 2003.
14. The Eclipse Framework. http://www.eclipse.org
15. M. Bastarrica et al. *Two Optimization Techniques for Component-Based Systems Deployment*, Proceedings of the Thirteenth International Conference on Software Engineering and Knowledge Engineering, pp. 153-162, 2001
16. IBM Corporation, *IBM Tivoli Software Homepage*, http://www.ibm.com/software/tivoli/
17. S. Chen, I. Gorton, A. Liu, and Y. Liu, *Performance Prediction of COTS Component-based Enterprise Applications*, CBSE5, Orlando, Florida, USA, May 2002.
18. A. Bertolino, R. Mirandola Software performance engineering of component-based systems. Proceedings of the Fourth Int. Workshop on Software and Performance, pp. 238 – 242, 2004.
19. J. Skene, W. Emmerich. *Model Driven Performance Analysis of Enterprise Information Systems*, In Proc. of the Int. Workshop on Test and Analysis of Component Based Systems, Warsaw, Poland, April, ENTCS vol. 82, num. 6, 2003.
20. V. Grassi. *Architecture-based Dependability Prediction for Service-oriented Computing*. In Proc. of WADS 2004,

21. V. Cortellessa, H. Singh, B. Cukic: *Early reliability assessment of UML based software models*. Proceedings of the Third Int. workshop on Software and Performance, pp. 302 – 309, ACM Press, 2002.
22. István Majzik, András Pataricza, and Andrea Bondavalli. *Stochastic dependability analysis of system architecture based on UML models*. In Rogerio de Lemos, Cristina Gacek, and Alexander Romanovsky, editors, Architecting Dependable Systems, volume LNCS-2677, pages 219-244. Springer, 2003.
23. István Majzik, András Pataricza, and Andrea Bondavalli. *Stochastic dependability analysis of system architecture based on UML models*. In Rogerio de Lemos, Cristina Gacek, and Alexander Romanovsky, editors, Architecting Dependable Systems, volume LNCS-2677, pages 219-244. Springer, 2003.

On Best-Effort and Dependability, Service-Orientation and Panacea

Aad van Moorsel

University of Newcastle upon Tyne,
School of Computing Science,
Newcastle, UK
aad.vanmoorsel@newcastle.ac.uk

Abstract. In this short position paper we argue that dependability technologies must be based on best-effort engineering principles, if they are to be useful for general-purpose enterprise and consumer IT. We will explain why 'best-effort based dependability' is not an oxymoron, but, instead, a necessity. We will also argue that service-orientation fits the best-effort engineering philosophy, and in that sense is part of the panacea for high-availability.[1]

1 Best-Effort and Dependability

We make the following observations:

- reliability and availability is increasingly important in enterprise, home and other networked IT systems
- these systems must rely on best-effort designs, which focus on scalability and simplicity over reliability and availability
- improving reliability or availability will always be at the cost of some other system property (such as scalability, ease-of-use, extensibility)

We conclude from these observations that traditional dependability solutions are typically ill-suited to respond to the demand for increased dependability, simply because they would sacrifice the system's scalability or other properties to too large an extend. Instead, to improve dependability properties of enterprise and consumer IT systems we require technologies based on advanced best-effort methods. Only then can dependability be improved without sacrificing other important system properties.

Advanced best-effort has the following implications on dependability design and research. First, we have to relax the stringency of the properties we are after. Instead of strict reliability guarantees, we need a probabilistic statement, or even a subjective statement about the objective. Subjective statements are possible, since they can be verified after the fact through user questionairres and

---

[1] This article is based on the author's position statement at ISAS 2005 during the panel discussion on 'Are Service-Oriented Architectures the Panacea for the High-Availability Challenge?'

M. Malek, E. Nette, and N. Suri (Eds.): ISAS 2005, LNCS 3694, pp. 99–101, 2005.

other information-gathering techniques. Secondly, as a consequence of dealing with less rigorous properties and designs, we have to be willing to rely on 'good engineering', trusting that sound design choices with dependability in mind, will indeed lead to improved dependability. We will not be able to proof *a priori* if dependability increases, and we therefore need to build up a body of process and engineering methodology that is known to 'do good' for a system's dependability. Finally, the fact that we can not define and proof system dependability *a priori* is compounded by the unpredictable usage, behaviour and attacks exhibited by modern and future IT systems. This implies that we must stress run-time measurement, based on which systems can be tuned, adapted and evolved in response to observed changes or unsatisfactory dependability.

2   Service-Orientation and Panacea

The above call for marrying dependability and best-effort can be seen in many lights. We argue it is absolutely necessary to combine these two elements to have impact on enterprise and customer IT systems. Obviously, this approach can not lead to systems that are as dependable as airplanes, nuclear power plants and other safety-critical systems. However, researching and practising best-effort dependability can lead to a body of processes, methodology, engineering guidelines and theory that can form a tool box for improved dependability for a wide range of enterprise and consumer systems. If anything comes close to a panacea for high-availability, it is marrying dependability and best-effort.

The best example to date of the power of best-effort dependability is TCP, but many more examples exist. In fact, all dependability work that is concerned with the problem of scalability, must compromise dependability to some extend (e.g., probabilistic protocols for consistency). This is a symptom of best-effort dependability. The call for autonomic computing solutions from the IT industry has a best effort flavour, although the vision statement in [2] poorly reflects the inherent limitations (in terms of achievable dependability) of introducing probabilistic and statistical approaches.

At the application level, we consider service-oriented architectures a piece of the solution [3]. Breaking up functionality, and limiting the dependencies between services, is part of the principles that one needs to follow to design dependable systems for enterprise and consumer IT. At the same time, such loose coupling implies that it is harder (or practically impossible) to implement certain dependability properties, such as exactly-once transactions. As with any form of best-effort dependability, adhering to service-orientation implies that one loses certain dependability options, but that is unavoidable when dealing with open, large-scale enterprise and consumer systems.

Note: Further discussion along above lines can be found in [3]. The call for dependable general-purpose systems by IBM can be found in the autonomic computing manifesto [2]. An interesting sample of recent research in the area of dependable, scalable systems can be found in the 'self-*' book [1].

References

1. O. Babaoglu, M. Jelasity, A. Montresor, C. Fetzer, S. Leonardi, A. van Moorsel and M. van Steen (Eds.), *Self-Star Properties in Complex Information Systems,* Springer, Lecture Notes in Computer Science vol. 3460, 2005.
2. P. Horn, "Autonomic Computing: IBM's Perspective on the State of Information Technology," *IBM,* USA, 2002. (Available at http://www.research.ibm.com.)
3. A. van Moorsel, "Grid, Management and Self-Management," *The Computer Journal,* British Computer Society, Oxford University Press, UK, vol. 48, no. 3, pp. 325–332, 2005.

Are Service-Oriented Architectures the Panacea for a High-Availability Challenge?

A Position Statement

Guido Laures

Hasso-Plattner-Institute for IT-Systems -Engineering at University of Potsdam,
Germany
Prof.-Dr.-Helmert-Strasse 2-3, 14482 Potsdam, Germany
guido.laures@hpi.uni-potsdam.de

Abstract. Service-oriented architectures (SOA) are based on a paradigm that aims at facilitating the management of business processes. Services are business relevant functionalities that are transparently provided by one or more applications. This simplified view on SOA has little in common with the view on technical and non-functional system properties from the high-availability research area.

After the presentation of service examples taken from various research approaches this position paper introduces a service layer model delimiting the notion of services. It is shown that specific layers of this model can help to determine and increase the availability of business processes. Furthermore the usability of service composition is questioned. The outlook in the last section of this position statement dares to predicts a shift from service-orientation to event- and business-orientation.

1 What Is a Service?

When discussing services or service-oriented architecture it is crucial to have a common terminology. Lately the term service has been overstrained and is frequently used at random. Asking someone 10 years ago to give an example for a service might have resulted in an answer like 'having one's hair cut'. During the last few years *service* has increasingly become a technical term. However, it still remains unclear what exactly a service is, even in the world of IT. The term is used for end-user services which are provided by a web site (e.g. ordering a book or home banking) as well as for the so called web services. The latter do not have a precise definition either. Related articles on service-oriented architecture state that amazon.com itself can be considered a web service [1]. Other approaches define a web service as a component with a SOAP-based access to its operations [2]. Especially in the telecommunication industry services are used for protocol-centric functions provided by networks. Recently Foster, et.al. [3] have introduced the grid services concept whereas Fremantle [4] and Krafzig, et.al. [5] adhere to enterprise service architectures. This ambiguity shows that a clustering of service types inside a layer model is mandatory.

M. Malek, E. Nette, and N. Suri (Eds.): ISAS 2005, LNCS 3694, pp. 102–106, 2005.

2   Service Layers

Different views can be applied to introduce layers into the SOA world. The first one is a view on the usage scenario of the services provided by an SOA; technical properties define the layers in the second view. These differentiating views on SOA are crucial to avoid confusing comparisons of usage-oriented and technology-based service clusters.

The most coarse-grained service type can be found in the enterprise service layer. Services on that layer can be accessed from outside the service providing entity. Therefore, they need to be very expressive and easy to use. In most cases a service from this layer triggers or is part of a business process containing outsourced activities. The second layer contains all intra-enterprise business processes that are provided as a service. Those mostly composite services are administered by a process execution engine holding the state of the process during its enactment. The layer below mainly contains stateless services which provide access to specific functionalities of a certain application domain. Access to data objects is granted by the data-centric services of the lowest layer. They can be invoked to create, read, update, or delete (CRUD) data objects. Figure 1 shows this layer model that is compatible to the model presented by Krafzig, et.al. [5].

Fig. 1. Usage scenario-oriented service layer model

Notice that this layering shows that the non-functional and architectural requirements for services strongly depend on their usage scenario. Services on layer 1, for instance, have stricter requirements for security, whereas services on the lower layers need to perform better. The technology that provides services also differs between the service layers. Standardized web service interfaces are needed for public access whereas the lower layers can be implemented using, for instance, a high performing message bus.

Thus, it makes sense to introduce a second layer model that is derived from a more technical view on SOA (see Fig. 2). In this model the transparency of the service providing technology decreases from top to bottom while the reusability of the services increases. Layer 1 contains the traditional services (e.g. having one's hair cut) that do not involve any computational resources. Frontend services are provided by a user interface (e.g. ordering a book at amazon.com). The web services world is represented by the third layer. Layers 4-7 contain functionalities that are sometimes also perceived as services. They range from component interfaces (CORBA, COM, RMI) to grid services or even CPU-cycles.

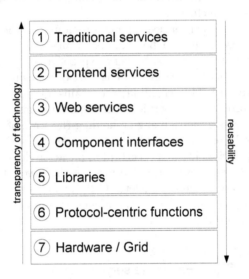

Fig. 2. Technology-oriented service layer model

Because of the immense differences between the various service layers in different views, perceiving everything as a service is misleading. If you name everything a service you can accordingly call every architecture a service-oriented architecture. Consequently, the specific benefits of a real SOA blur. Thus, a layering like the aforementioned helps to define clear conceptual differences and to expose its value-adding concepts. Therefore, the following chapters use the term *service* for services in usage scenario layers 1 and 2 and technological layer 3. In doing so, I follow the initial concepts on service-orientation as published in [6].

3   SOA and High-Availability

When investigating the aforementioned usage layer 1-2 and the technological service layer 3 (SOAP-based services) analysing high-availability is difficult. The reason for this is that these layers do not deal with the classical hardware- or protocol-centric availability challenges. Instead they rather deal with the availability of processes.

In a classical, non service-oriented enterprise architecture applications offer their specific functionalities via remote interfaces. The different types of such interfaces range from simple batch file transfer mechanisms like FTP, to component remote interfaces like COM or CORBA, up to SOAP-enabled web services. The activity flow of an application-spanning business process is implicitly hard coded inside the application logic and its resulting inter-application communication. As every remote interface provides functionality that can be used in a large number of different business processes it is not possible to determine which business process is currently conducted by a certain interface. Thus, all examinations on non-functional properties of remote interfaces are application-centric and unaware of any business context. Knowing the availability of a certain remote interface does not provide any information on the availability of the business processes using it.

Service-orientation aims at providing business process oriented functionalities in a unified fashion. Service enactment components such as workflow engines control the invocation of specific services according to a well-defined process description. This allows to determine business process availability based on the availability of the services used by this process. Additionally, the automatic identification of business processes affected by the availability of application domain services becomes possible. The real-time retrieval of this kind of information has become a key differentiating factor in many industries (esp. telecommunication). Because many service level agreements (SLA) nowadays are rather business- than technology-oriented it is vital to find mechanisms to prove business process availabilities. This is where service-orientation kicks in.

4   Service Composition vs. Traditional Programming

When dealing with business processes, service composition is often highlighted as one of the core benefits of an SOA. The concept of service compositions is to combine existing services to implement new applications and processes. However, the aforementioned technology-based service layer model shows that the reusability of services decreases the more coarse-grained a service is designed. Thus, service compositions have to be based on the low-level service layers 4-7. However, the services on those layers provide access to functionalities that are often more technologically driven than those on the layers above. A composition of such services is not just a selection of an appropriate sequence of service invocations but rather evolves into a complex workflow that is often just as complex as a classical program.

Standards like BPEL [7] have become so complex that they can only be used by process technology experts but not by experts in the industrial domain. Consequently, composing is nothing less but another kind of programming. The (semi-)automation of service composition planning [8] might be a promising approach to solve this problem.

5   Conclusion and Outlook

This short position statement refined the notion of services by introducing layer models. These layers help to concentrate on the usage scenarios and technologies where SOA is applicable. It showed that the layers of services and the layer on which high-availability calculations base differ. Thus, the contribution of SOA to the high-availability approaches is little. As a conclusion service-orientation helps to administrate IT-architectures but it does not consider availability issues by default. On a process level, however, service-orientation can help to measure and consequently optimise the availability of business cases.

Because the service notion is overstrained I expect an oversaturation in research and business followed by decreasing interest in service-orientation. Even though the paradigm can be leveraged to solve actual business problems it has to be enhanced by event-driven approaches and a stronger orientation on business needs. During this clean-up phase in the world of service-orientation some hot topics like service semantics or services grids are yet to prove capable.

References

1. McIlraith, S.A., Martin, D.L.: Bringing semantics to web services. IEEE Intelligent Systems **18** (2003) 90–93
2. W3C: Web Services Architecture. (2004) `http://www.w3.org/TR/ws-arch/`.
3. Foster, I., Kesselman, C., Nick, J., Tuecke, S.: The physiology of the grid: An open grid services architecture for distributed systems integration (2002)
4. Paul Fremantle, S.W., Khalaf, R.: Enterprise services (2002)
5. Krafzig, D., Banke, K., Slama, D.: Enterprise SOA. Prentice Hall (2004)
6. IBM: Web Services architecture overview. (2000)
   `http://www-106.ibm.com/developerworks/webservices/library/w-ovr/`.
7. Organization for the Advancement of Structured Information Standards: Web Services Business Process Execution Language (WS-BPEL). (2004) `http://www.oasis-open.org/committees/tc_home.php?wg_abbrev=wsbpel`.
8. Sirin, E., Hendler, J., Parsia, B.: Semi-automatic composition of web services using semantic descriptions. In: Web Services: Modeling, Architecture and Infrastructure workshop in ICEIS 2003. (2003)

Modeling User-Perceived Service Availability

Dazhi Wang[1] and Kishor S. Trivedi[2]

Duke University, Durham, NC, USA 27708
wangdz@cs.duke.edu, kst@ee.duke.edu

Abstract. Service availability is an important consideration when carriers deploy new, packet-based services. In this paper we define the service availability based on user behavior, and derive formulas to compute service availability starting with the user behavior model and the system model. To automatically generate high fidelity user and system models, we use Stochastic Reward Nets (SRNs) and demonstrate how to combine the user SRN model and the system SRN model to analyze the service availability. We apply our approach to an SAF compliant media gateway controller (MGC) architecture in VoIP system. By building and numerically solving the combined SRN model of the MGC and the user, we compute the service availability, and evaluate various factors that influence it.

1 Introduction

Today the fast development of new technologies have enabled a variety of new services for voice, data and multimedia. Ensuring high service availability is important as users become more dependant on these services to conduct their everyday activities. To achieve high service availability with lower cost, the Service Availability Forum (SAF) [1] was created by a group of premier communications and computing companies. Its goal is to create and promote open standards that will build the foundation for on-demand, uninterrupted network services delivered over packet-switched communication networks, and help make these services as dependable as those delivered through traditional Public Switched Telephone Networks (PSTN). By conforming to the interface specifications, the hardware producers can create open, COTS building blocks that have a higher reusability and a larger market; while software vendors can reduce the time-to-market and development cost for highly available software with enhanced portability and integration capabilities.

As the SAF develops open standards to help meet end-user expectations for high availability services, the need for the quantification of service availability becomes evident. The modular architecture of SAF-compliant systems enables the succinct analytical modeling of service availability and performance, which allows users to do "what if" analysis by combining different system configurations, policies and building blocks to meet various availability and performance requirements. Previous research mainly focused on the traditional availability measures such as point availability, interval availability or steady-state availability, whose definitions can be found in [3] [13] [15]. These measures are primarily

M. Malek, E. Nette, and N. Suri (Eds.): ISAS 2005, LNCS 3694, pp. 107–122, 2005.

from the system's point of view, representing a fraction of a time period the system is up or the probability that the system is up at certain time point. However service providers are more concerned with the user-perceived service availability, which is a function of system hardware resource availability, software resources and user behavior. Therefore it is important to characterize the user behavior in order to model service availability. During the user interaction with the system, the user often submits multiple requests to achieve certain goals, each of the requests may require different resources in the system. These characteristics can be summarized as follows:

- Not all resources in the system are required to be up at every point in time. This is determined by the complex nature of the modern systems, and motivates us to focus on service availability instead of the system availability.
- A resource is required to be up only during the time periods when the user requests this resource. This characteristic determines that our service availability analysis must be user-centric.
- A resource may be required to be up for multiple time periods during the user interaction with the system. Due to this requirement, traditional point availability measures can not be applied in service availability analysis. Instead, joint availability [3] or interval reliability [2] may be well suited for this purpose.

Based on these observations, the service availability analysis needs to take into account the details of user behavior, and it should adopt a dynamic view of system up/down states (when needed, as long as needed, as many times as needed). We interpret the **user-perceived service availability**[1] as follows:

During a user interaction (session) with the system, the user issues multiple requests at different time points for different system resources. The unavailability of requested resource will cause the request to fail. The service availability is the probability that all requests are successfully satisfied during the user session.

Some research effort related to service availability can be found in the literature, such as the user-perceived availability [12] [18] or the task-oriented reliability/availability [7] [8] [14]. However these efforts did not fully explore the service characteristics provided by the modern complex systems and experienced by the end users. In order to model the service characteristics as mentioned above, we borrow the user behavior graph approach [4] that was originally used for workload synthesis, and derive the service availability starting from the user behavior model and the system availability model. The rest of the paper is organized as follows: Section 2 derives service availability formulas from user and system models, and proposes the use of Stochastic Reward Nets (SRNs) [6] [16] to automatically combine user and system models to compute the service availability. Section 3 applies our service availability modeling technique

---

[1] It can be argued that this measure is better interpreted as user-perceived service reliability, but to conform to the standard practice, we use the term user-perceived service availability.

to an SAF-compliant media gateway controller (MGC) architecture. Section 4 gives the numerical results of service availability modeled in Section 3, and analyzes various factors that influence the service availability. Section 5 presents the summary and conclusions.

2 Service Availability Modeling

2.1 User Behavior Graph (UBG)

To model the service availability we first introduce the concept of user behavior graph [4] [5]. The user behavior graph is composed of a set of nodes and arcs. Each node indicates a certain type of request issued by the user. A transition implies a new request is issued, and an arc to node j represents the newly issued request is of type j. Each arc is attached a probability that the next issued request is of the type this arc points to. The sojourn time in each node is the sum of the time the system takes to process the request and the think time after the user receives the response. Figure 1(a) shows an example user behavior graph

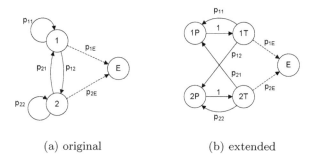

(a) original   (b) extended

Fig. 1. Example user behavior graph with two types of requests

with two types of requests. For each node $i = 1, 2$, $p_{ij}, j = 1, 2$ is the probability that the user will issue a request of type j after the current request of type i. There is one absorbing node E representing session end. The probability that the user will end the session after issuing a request of type i is p_{iE}. Since the occurrence of a failure during the request processing will influence the user-perceived service availability while a failure during the user think time will not, we extend the graph by splitting each node into two nodes for our purpose: one representing the request processing state, and the other representing the user think state. Figure 1(a) then becomes Fig. 1(b). Given a user behavior graph that describes a user session, the user-perceived service availability is defined as the probability that all requests in the session are successfully completed when the user enters E state.

2.2 User-Perceived Service Availability Derivation

In this section we introduce the computation of user-perceived service availability, starting from simple examples.

Single Task Service Availability. Given a two-state system model with constant failure rate λ and repair rate μ, the system dependability measures are shown in Table 1.

Table 1. Dependability measures of the two-state system

Reliability	Instantaneous Avail.	Steady-state Avail.	Interval Relia.
$R(t) = e^{-\lambda t}$	$A(t) = \frac{\mu}{\lambda+\mu} + \frac{\lambda}{\lambda+\mu} \cdot e^{-(\lambda+\mu)t}$	$A = \frac{\mu}{\lambda+\mu}$	$R_I(t,x) = A(t) \cdot e^{-\lambda x}$

Now we consider the user-perceived service availability of a single user task running on the two-state system. The system model is shown in Fig. 2(a) with constant failure rate λ and repair rate μ, and the user model is shown in Fig. 2(b) which contains only one request. Assume the initial service accessibility is

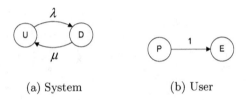

(a) System   (b) User

Fig. 2. System and user models for single task reliability

u_0, then the user perceived service availability can be written as:

$$SA = u_0 \cdot \int_0^\infty Pr(\text{ system is up during } [0, y]) \cdot Pr(\text{request completes at time } y)$$

$$= u_0 \cdot \int_0^\infty R(y) \cdot dF_P(y)$$

If the request processing time is deterministic, *i.e.*, the sojourn time distribution function $F_P(y)$ in user state P is 0 when $y < x$, and 1 otherwise. Then

$$SA = u_0 \cdot \int_0^\infty R(y) \cdot dF_P(y) = u_0 \cdot e^{-\lambda x}.$$

If the request processing time is exponentially distributed with rate λ_P, *i.e.*, $F_P(y) = 1 - e^{-\lambda_P y}$, then SA can be written as

$$SA = u_0 \cdot \int_0^\infty R(y) dF_P(y) = u_0 \cdot \int_0^\infty e^{-\lambda y} \cdot \lambda_P \cdot e^{-\lambda_P \cdot y} dy = u_0 \cdot \frac{\lambda_P}{\lambda + \lambda_P}.$$

Service Availability for UBG with Loops. In this example we extend the user behavior graph of the previous example to allow multiple requests in the user session. The system model and the UBG are shown in Fig. 3. In the UBG

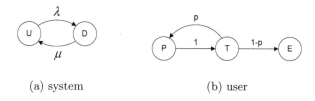

(a) system   (b) user

Fig. 3. System model and UBG with loop

of Fig. 3(b), after leaving the request processing state P, the user will enter thinking state T. And upon leaving T state, with probability p the user will send another request, and with probability $1 - p$ the user will end the session. We assume the request processing time distribution $F_P(t)$ and the user thinking time distribution $F_T(t)$ are both exponentially distributed with rate λ_P and λ_T, respectively. The system reliability $R(t)$ and instantaneous availability $A(t)$ are as in Table 1. We define the service continuity

$$C = \int_0^\infty R(t)dF_P(t) = \frac{\lambda_P}{\lambda + \lambda_P},$$

and the service accessibility

$$D = \int_0^\infty A(t)dF_T(t) = \frac{\mu}{\lambda + \mu} + \frac{\lambda}{\lambda + \mu} \cdot \frac{\lambda_T}{\lambda + \mu + \lambda_T}.$$

Then C is the probability that a request can complete given it is initiated in the system, and D is the probability that the user can successfully initiate a request upon leaving the thinking state. Given the initial service accessibility u_0, the probability S_i that the user ends the session after i successful requests can be written as:

$$S_i = u_0 \cdot C^i (pD)^{i-1} \cdot (1 - p) = u_0 (1 - p) C \cdot (p \cdot C \cdot D)^{i-1}$$

And the user-perceived service availability is:

$$SA = \sum_{i=1}^\infty S_i = \frac{u_0 \cdot (1 - p) \cdot C}{1 - p \cdot C \cdot D} = \frac{u_0 \cdot \frac{\lambda_P}{\lambda + \lambda_P} \cdot (1 - p)}{1 - p \cdot [\frac{\lambda_P}{\lambda + \lambda_P} \cdot (\frac{\mu}{\lambda + \mu} + \frac{\lambda}{\lambda + \mu} \cdot \frac{\lambda_T}{\lambda + \mu + \lambda_T})]}$$

Service Availability for General System and User Models. To derive the formulas for service availability computation, we generalize the simple system and user models in previous examples as follows:

- The UBG is a general discrete time Markov chain (DTMC) with $m+1$ states, where the $(m+1)$th state is the absorbing session end state E. The state space $\Omega_U = \{1, 2, ..., m, E\}$. Let $\mathbf{P} = (p_{ij})_{m \times m} = (\mathbf{P}_1, \mathbf{P}_2, ..., \mathbf{P}_m)$ be the transition probability matrix for the first m states, and $\mathbf{q} = (p_{1E}, p_{2E}, ..., p_{mE})$. The initial probability vector when the session begins is $\mathbf{v}_0 = (v_1, v_2, ..., v_m)$. The sojourn time of each state i is arbitrarily distributed with distribution function $F_i(t)$. For a request processing state the sojourn time is the request completion time, while for an user think state the sojourn time is the user think time. Matrix $\mathbf{P}$ and the set of sojourn time distributions $\{F_i(t)\}$ define an independent Semi-Markov Process (SMP).
- The system availability model (SAM) is a continuous time Markov chain (CTMC) with state space $\Omega_S = \{1, 2, ..., n\}$ and generator matrix $\mathbf{Q}$. The initial probability vector of SAM at the beginning of the session is $\boldsymbol{\pi}_0 = (\pi_1, \pi_2, ..., \pi_n)$. Each request processing state i in the user model needs different resources to be up in the system, which can be translated to requiring the system model to be in a certain subset of states. Therefore each user state i has its own definition of system up states $R_i \subseteq \Omega_S$ and down states $\Omega_S - R_i$ (For user think state all system states are up states). By removing the outgoing transitions from system down states in $\Omega_S - R_i$, $\mathbf{Q}$ becomes $\mathbf{Q}_i$ for each user state i. We define $\mathbf{D_i} = (d_{kj})_{n \times n}$, where $d_{kj} = 0$ for $k \neq j$, $d_{kk} = 1$ if $k \in R_i$, otherwise $d_{kk} = 0$. Then for a vector $\mathbf{v}$, $\mathbf{v} \cdot \mathbf{D_i}$ is the vector where each entry with an index is not in R_i is set to 0. For all the user think states, $\mathbf{D}_i = \mathbf{I}_{n \times n}$, which means no resource is needed and the system model can be in any state.

In order to derive the service availability from the generalized user and system models, we define $\boldsymbol{\Pi}(k) = (\pi_{ij}(k))_{m \times n}$, and $\boldsymbol{\Pi}_i(k) = (\pi_{i1}(k), \pi_{i2}(k), ..., \pi_{in}(k))$, where $\pi_{ij}(k)$ is the probability that the user model is in state i and the system model is in state j right before the kth transition in the user model.

With the explanations above we derive the service availability as follows:

$$\text{For } k = 1, \boldsymbol{\Pi}_i(1) = v_i \cdot \boldsymbol{\pi}_0 \mathbf{D}_i \cdot \int_0^\infty e^{\mathbf{Q}_i \cdot t} dF_i(t) = v_i \cdot \boldsymbol{\pi}_0 \mathbf{D}_i \cdot \mathbf{H}_i$$

where v_i is the probability that the user model is initially in state i, $\boldsymbol{\pi}_0 \mathbf{D}_i$ keeps only the entries in $\boldsymbol{\pi}_0$ that correspond to states in R_i, $\mathbf{H}_i = \int_0^\infty e^{\mathbf{Q}_i \cdot t} dF_i(t) = (h_{jk})_{n \times n}$, h_{jk} is the probability that given the initial state j, the system is in state k at the time when the first transition occurs in the user model. And the probability that the session will successfully end after one transition in the user model is: $A(1) = \mathbf{q} \cdot \boldsymbol{\Pi}(1) \cdot \mathbf{1}^\mathbf{T}$.

Given $\boldsymbol{\Pi}(k-1)$, $\boldsymbol{\Pi}(k)$ can be computed using (1) shown below. Here we only present the derived formula due to space limitations.

$$\boldsymbol{\Pi}(k) = \begin{pmatrix} \mathbf{P}_1^\mathbf{T} \cdot \boldsymbol{\Pi}(k-1) \cdot \mathbf{D}_1 \mathbf{H}_1 \\ \mathbf{P}_2^\mathbf{T} \cdot \boldsymbol{\Pi}(k-1) \cdot \mathbf{D}_2 \mathbf{H}_2 \\ \vdots \\ \mathbf{P}_m^\mathbf{T} \cdot \boldsymbol{\Pi}(k-1) \cdot \mathbf{D}_m \mathbf{H}_m \end{pmatrix} \tag{1}$$

The probability that the session will successfully end after k transitions in the user model is:

$$A(k) = \mathbf{q} \cdot \mathbf{\Pi}(k) \cdot \mathbf{1^T}$$

From the above deductions, the total service availability is

$$SA = \sum_{k=1}^{\infty} A(k) = \mathbf{q} \cdot \sum_{k=1}^{\infty} \mathbf{\Pi}(k) \cdot \mathbf{1^T}$$

2.3 SRN Model of User Behavior

In order to apply the formulas above to the computation of service availability, we need to construct the user model and the system model. However as the modern systems become more and more complex, it is often infeasible to manually construct a high fidelity system model. To solve this problem we resort to Stochastic Reward Nets (SRNs) [6], which is a higher level formalism based on Stochastic Petri Nets (SPN) [9] [17]. Since SRNs can only deal with exponentially-distributed timed transitions, as a compromise we assume the sojourn times in the user model and the transition times in the system model are exponentially distributed.

By assuming each state has exponentially distributed sojourn time, we can build the stochastic Reward Net (SRN) model for the user behavior graph as shown in Fig. 4. A token in $P_{iP}, i = 1, 2$ means a request of type i is being

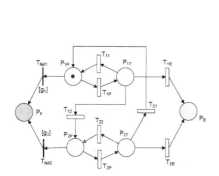

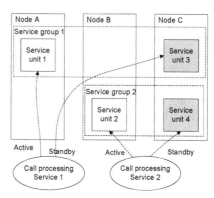

Fig. 4. SRN model for the user behavior graph

Fig. 5. Two services with active/standby on three cluster nodes

processed. A token in $P_{iT}, i = 1, 2$ means the user is thinking after receiving the response for request type i. The firing of $T_{ij}, i = 1$ or 2, $j = 1$ or 2 represents the issuance of the current request is of type i and that the next issued request is of type j. The firing of $T_{iE}, i = 1$ or 2 puts the token into the absorbing place P_E, representing successful session end. There are two immediate transitions T_{fail1}

and T_{fail2} with guard functions g_1 and g_2. The return values of g_1 and g_2 are determined by markings in the system model: g_1 will return true if the system resources required to process requests of type 1 are unavailable; g_2 will return true if the system resources for request type 2 are unavailable. If there is a token in P_{1P} (meaning a request of type 1 is being processed) and g_1 returns true, T_{fail1} will fire and put the token in the absorbing place P_F, representing service failure. And similarly for T_{fail2}. Combining the user SRN model and the system SRN model, the service availability is the probability that there is a token in the absorbing place P_E.

3   Service Availability of an SAF Compliant MGC

In this section we apply the service availability modeling technique to an SAF compliant media gateway controller (MGC) for voice over IP (VoIP).

3.1   Basic Architecture

The media gateway controller is hosted on three identical nodes in a computer cluster. The MGC application is SAF AIS-compliant and the availability is managed by the SAF AIS Availability Management Framework (AMF). The main function for the media gateway controller is for call processing, which is required during the call setup, teardown, or when the user invokes additional call features in the middle of a call. There are two application servers running on the cluster for call processing, each in charge of processing different call features. Process replication is adopted as the mechanism to provide application level software fault tolerance [10] [11]. These service processes run on top of the Service Availability Forum Middleware. The system model is shown in Fig. 5. Each service application has one service instance, and each service instance is assigned to two service units in different cluster nodes: one service unit is active and the other is standby. Either node A or node B has 1 service unit acting as the primary for call processing service 1 or call processing service 2. Node C has two service units acting as a shared standby for both services.

  The service unit may fail due to either software fault or hardware fault. The fault will be detected by the health monitoring mechanisms in the AMF. After the fault has been detected, recovery strategies are adopted to tolerate the fault. Here we assume that for software faults the AMF tries several levels of recovery actions: 1) component restart, which is fast and has little or no impact on the application; 2) switchover that switches the service to the standby service unit, in the mean time the faulty node is restarted; 3)manual repair of the cluster node if automatic restart of the previous level cannot recover the fault. For hardware faults, we assume recovery actions 2 and 3 are adopted by the AMF. After the fault is detected the AMF tries to recover the fault using from lower level to higher level recovery strategies, each level requiring more time to execute than the previous level.

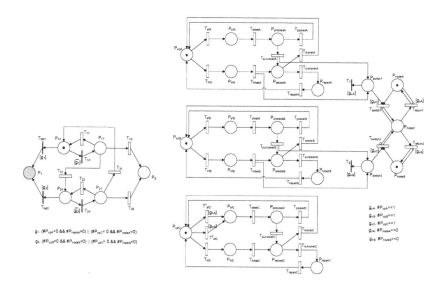

Fig. 6. SRN model for the service

3.2   SRN Model for the Service Availability

Figure 6 shows the SRN combining together the user and system model. The right part of Fig. 6 shows the SRN for the system model. When the service is first started, all three nodes are up, which is represented by a token in P_{upA}, P_{upB} and P_{upC}, respectively. Service 1 is running on node A while service 2 is running on node B. For node A, there can be either service process fault with rate λ_{sfA}, or the hardware fault with rate λ_{hfA}. The fault is detected by the health checking mechanisms in the Availability Management Framework, which is represented by transitions T_{sfdetA} and T_{hfdetA}.

For node A, the recovery actions are represented in the SRN model by tokens in place $P_{processA}$, $P_{rebootA}$, or $P_{repairA}$. The firing of $T_{pcoverA}$, $T_{rcoverA}$ or $T_{repairA}$ means the fault is successfully recovered by component restart, node restart, or manual repair, while firing of $T_{puncoverA}$ or $T_{runcoverA}$ means the fault is not covered by the corresponding recovery mechanism and escalated to a higher level recovery action. We assume manual repair can always fix the fault.

Node B has similar behavior as node A, with possibly different rates and coverage factors for these events. For node C, the failure behavior of its components is slightly different. Since they act as standby components, we assume software faults will not occur on node C as long as neither service is switched onto it. When a switchover occurs, however, either T_{sfC}^1 (guarded by g_{nA}) or T_{sfC}^2 (guarded by g_{nB}) or both will be enabled depending on the switched service(s) on C, and software failure could occur. The corresponding transition will be disabled again after the service is switched back.

A token in place P_{nodeA}, P_{nodeB} or P_{nodeC} means a service process is hosted on the corresponding node A, B, or C. Initially there is one token in P_{nodeA} and one in P_{nodeB}. For node A, when the reboot action is taken, indicated by the firing of either transition $T_{puncoverA}$ or T_{hfdetA}, a token is put into place $P_{switch1}$ in addition to place $P_{rebootA}$ to start the switchover process. If node C is up (guarded by function g_{uC}) and the service token is in P_{nodeA}, $T_{switch1}$ will be enabled to switch to service process to the standby node C. When node A is recovered from the fault, $T_{return1}$ will be enabled (guarded by function g_{uA}) and the firing of $T_{return1}$ indicates that the service is switched back to node A. The switchover of the service process on node B can be similarly done. As long as node C acts as a primary for either call processing service 1 or service 2 (represented by a token in P_{nodeC}), it can no longer be the primary for the other service. This is guaranteed by the inhibitor arcs from P_{nodeC} to $T_{switch1}$ and $T_{switch2}$. The guard functions in the system model are shown in Table 2.

Table 2. Guard functions for the system model in Fig. 6

guard function	true condition	description
g_{uA}	$\#P_{upA} = 1$	node A is up
g_{uB}	$\#P_{upB} = 1$	node B is up
g_{uC}	$\#P_{upC} = 1$	node C is up
g_{nA}	$\#P_{nodeA} = 0$	primary for service 1 is on node C
g_{nB}	$\#P_{nodeB} = 0$	primary for service 2 is on node C

When the user makes a call, he may send multiple call processing requests, such as call setup, call feature invocation, or call teardown, to different application servers in the media gateway controller. The user behavior during this procedure is described by the user behavior graph in Fig. 1(b): state $iP, i = 1, 2$ is the call processing state which requires application server i to be up; state $iT, i = 1, 2$ is the user talking state in which neither server is required. When the user is in state $1T$, with probability p_{11} the user will invoke a call feature that needs processing in application server 1, with probability p_{12} the user will invoke a call feature that needs processing in application server 2, with probability p_{1E} the user will terminate the call without requesting additional call features. And similarly for state $2T$. The left part of Fig. 6 shows the corresponding SRN model which is similar to Fig. 4.

For simplicity and consistency, we still use the 'request processing state' and 'user think state' to refer to the 'call feature processing state' and the 'user talking state' in our example user model. We assume the call setup is handled by application server 1, therefore there is a token in place P_{1P} when the call session first starts. The guard function g_1 returns true when application server 1 is unavailable and guard function g_2 returns true when application server 2 is unavailable. From the system model in Fig. 6, g_1 and g_2 can be expressed in Table 3.

Table 3. Guard functions for the user model in Fig. 6

guard functions	true condition
g_1	$(\#P_{upA} = 0 \text{ and } \#P_{nodeA} > 0)$ or $(\#P_{upC} = 0 \text{ and } \#P_{nodeA} = 0)$
g_2	$(\#P_{upB} = 0 \text{ and } \#P_{nodeB} > 0)$ or $(\#P_{upC} = 0 \text{ and } \#P_{nodeB} = 0)$

We assume the call processing request is served quickly enough that the processing time can be neglected. Therefore the timed transitions T_{1P} and T_{2P} in Fig. 4 are replaced by immediate transitions in Fig. 6, and are guarded by $\bar{g}_1$ and $\bar{g}_2$, respectively.

4   Numerical Results

In this section we evaluate the service availability under various input parameters using the SRN model of the previous section. The default parameter values used in the model are shown in Table 4. For comparison purposes, we also draw

Table 4. Default parameters used in the model

Parameter	Default value	Description
MTTF_{sft}	336 hours	software MTTF for each service unit
MTTF_{hd}	672 hours	hardware MTTF for each service unit
f_{detect}	7200 hour^{-1}	detection rate for sw/hw faults
c_{proc}	0.95	coverage factor for component restart
c_{reboot}	0.9	coverage factor for node restart
MTTR_{proc}	10 seconds	mean time for component restart
MTTR_{reboot}	3 minutes	mean time for node restart
MTTR_{repair}	8 hours	mean time for manual repair
S	5 minutes	user think time in state $iT, i = 1, 2$
p_{i1}	0.4	value for p_{i1} in the user model, $i = 1, 2$
p_{i2}	0.3	value for p_{i2} in the user model, $i = 1, 2$
p_{iE}	0.3	value for p_{iE} in the user model, $i = 1, 2$

the curves for steady-state system availability and the lower bound of service availability. The former is computed as the probability that both servers are up in steady state, and corresponds to the 'System' curve in the figures in this section. The latter corresponds to the 'SA-LB' curve and can be viewed as the service availability with infinite user think time. It is computed by hierarchically solving the user model and the system model, i.e., first use the system model to acquire p_i $i = 1, 2$, the steady-state probability that server i is up, then assign p_i as the transition probability for T_{iP} and $1 - p_i$ for T_{faili} in the user model, and solve the user model to get the probability that the token enters the absorbing place P_E.

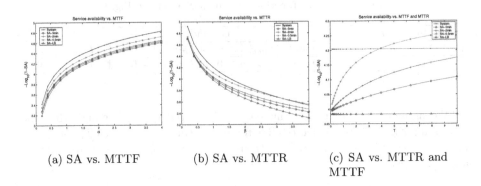

(a) SA vs. MTTF   (b) SA vs. MTTR   (c) SA vs. MTTR and
                                                      MTTF

Fig. 7. Service availability under different MTTFs and MTTRs

4.1   Service Availability vs. MTTF and MTTR

We first study the impact of system MTTF and MTTR on service availability. In Fig. 7, three service availability curves are drawn with 0.5, 2 and 5 minutes user think time. Figure 7(a) shows the service availability under different system MTTF, where the default $MTTF_{sft}$ and $MTTF_{hd}$ values in Table 4 are multiplied by a factor α ranging from 0.2 to 4. The x-axis shows the values for α, and y-axis shows $-\log_{10}(1 - SA)$, which computes the number of 9s of the service availability. Figure 7(b) varies the MTTR parameters in Table 4 ($MTTR_{proc}$, $MTTR_{reboot}$, $MTTR_{repair}$, $MTTR_{switch}$) by multiplying them with factor β ranging from 0.2 to 4, and the x-axis shows β values. Figure 7(c) multiplies both MTTF and MTTR parameters by factor γ ranging from 0.1 to 10 while keeping the system availability constant. From these figures it is obvious to see that the service availability increases with MTTF and decreases with MTTR. It is also worth noticing from Fig. 7(c) that while maintaining the system availability unchanged, longer MTTF and MTTR actually improve the service availability within the context of this paper. This is because the service availability can be viewed as the joint availability [3] at the request-arrival times during the user session, and the joint availability is higher when the system has longer MTTF and MTTR. From Fig. 7(a) and 7(b) we can see that the improvement is mainly caused by the longer MTTR, since the distance between SA curves and SA lower bound curve remains the same as MTTF increases, while the distance becomes larger as MTTR increases. Using the default parameters in Table 4, the service availability is 0.999903, the system availability is 0.999938, and the SA lower bound is 0.999896.

4.2   Service Availability vs. Coverage Factors

Figure 8 shows the impact of coverage factors on the service availability. We change the c_p and c_r values in Table 4 by adding θ, i.e., $c_p = 0.95 + \theta$, $c_r = 0.9 + \theta$, where θ ranges from -0.1 to 0.04. The x-axis shows the θ value and y-axis shows

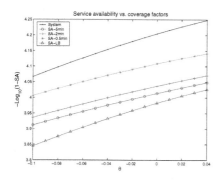

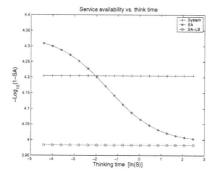

Fig. 8. SA vs. coverage factors   **Fig. 9.** SA vs. user think time

$\log_{10}(1 - SA)$. As seen from the figure, the service availability increases with coverage factors, and as the coverage factors increase, the SA curves become close to the SA-LB curve. The reason is that increasing coverage factors avoids higher level recovery actions which cost longer time than lower level recovery. Therefore it is equivalent to reducing the MTTR. As we observed from Sect. 4, smaller MTTR reduces the service availability improvement.

4.3   Service Availability vs. User Think Time

Figure 9 shows the service availability under different user think time S. S is varied from 1 second to 12 minutes. The x-axis is $\ln(S)$, and y-axis is $\log_{10}(1 - SA)$. As shown in the figure, the service availability drops as the user think time increases. When S goes to infinity, the SA will converge to the lower bound of SA. It can also be seen that when user think time is small, the service availability is even higher than the system availability. This can be explained by the following two factors: first, the joint availability for Markov models $JA(t, x)$ will increase and converge to the instantaneous availability $A(t)$ as x decreases to 0. Shorter user think time leads to smaller x. As we mentioned in Sect. 4, the service availability is similar to the joint availability in this aspect. Second, the system availability is the probability that both application servers are up. While in the user behavior model the user may only access one application server during the whole session. The probability that one server is up is larger than the system availability. Therefore the service availability can be greater than the system availability, although it may require parts of the system to be up for multiple times.

4.4   Service Availability vs. Session End Probability

Figure 10 shows the impact of session end probability to the service availability. In this figure we set $p_{1E} = p_{2E}$ and vary them from 0.1 to 0.9. As shown in the figure, the service availability increases with the session end probability. The

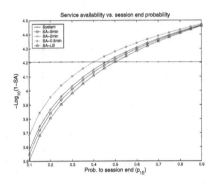
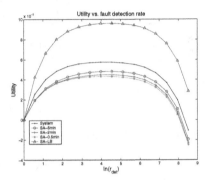

Fig. 10. SA vs. session end prob   **Fig. 11.** Utility vs. detection rate

reason that the service availability is greater than the system availability is the same as in Sect. 4.3: as the session end probability increases, the number of requests in the session is reduced. Since each request requires part of the system to be up which has a higher probability than the system availability, the service availability may exceed the system availability when p_{1E} and p_{2E} are large.

4.5   Impact of Detection Rate

When a fault occurs, it is detected by the Availability Management Framework using periodic health checking mechanisms. The higher the checking rate, the faster the fault can be detected, on the other hand, the more the system performance is degraded. In this section we study the tradeoff between service availability and performance under various fault detection rates. To do the analysis we make the following assumptions:

1. the system can perform at most X health checks per hour.
2. health checking at rate x utilizes $\frac{x}{X} \times 100\%$ computing power, thus reducing $\frac{x}{X} \times 100\%$ of the maximum throughput.
3. the reward rate for availability gain is C_1, and the penalty rate for performance loss is C_2.

Based on the assumptions above, we can write the utility function $U(r)$ as follows:

$$U(r) = C_1 * [SA(r) - SA(r_0)] - C_2 * [DE(r) - DE(r_0)]$$
$$= C_1 * [SA(r) - SA(r_0)] - C_2 * (r - r_0)/X$$

where $SA(r)$ is the service availability with detection rate r and $DE(r)$ is the percentage of throughput degradation. r_0 is the minimum detection rate used in the model.

We choose $C_1 = C_2 = 1$, $X = 7.2 \times 10^5 h^{-1}$, and vary the detection rate r from 1 per hour to 7200 per hour. Of course the parameters can be changed for different utility functions. Figure 11 shows the utility function with various detection rates. The x-axis is the natural logarithm of the detection rate r, y-axis is the utility function. From the figure it can be seen that when the detection rate is low, it is better to sacrifice a small percentage of performance to gain a much higher service availability; while when the detection rate is high, it is better to sacrifice availability for performance.

5   Conclusions

In this paper we evaluate user-perceived service availability by taking into consideration the user behavior model. The formulas for efficient service availability computation are derived for CTMC-based system model and for DTMC-based user model which allow arbitrary sojourn time distributions in each state. In order to enhance the model to characterize various aspects of user/system behavior, and yet avoiding the manual construction of large state-space models, we advocate the use of a higher level modelling formalism based on Stochastic Reward Nets (SRN). We have proposed an approach to combine the SRN-based user model and system model for service availability computation, and applied this approach to an example SAF-compliant MGC architecture.

We numerically solve the combined SRN model to compute the service availability. Various factors that influence the service availability are identified, such as mean time to failure, mean time to repair, fault coverage factors, user think times, session end probability, and fault detection rate. Service availability under these factors is evaluated and compared to system availability and service availability lower bound. The numerical results show that while keeping the system availability constant, longer MTTF and MTTR can increase the service availability in the context of this paper. On the other hand longer user think time can decrease the service availability. The tradeoff between availability improvement and performance degradation is also analyzed using the SRN model.

Acknowledgements

We thank Veena Mendiratta of Lucent and David Penkler of HP for many insightful discussions on VoIP and SAF-compliant architecture modeling. We thank Service Availability Forum for their funding to support this research.

References

1. http://www.saforum.org, last checked on May 24th, 2005.
2. R. E. Barlow and F. Proschan. *Mathematical Theory of Reliability*. New York: John Wiley and Sons, 1965.
3. A. Birolini.   *Quality and Reliability of Technical Systems: Theory-Practice-Management*. Springer-Verlag, Berlin, 1998.
4. M. Calzarossa and D. Ferrari. A sensitivity study of the clustering approach to workload modeling. *Performance Evaluation*, 6:25–33, 1986.

5. M. Calzarossa, R. A. Marie, and K. S. Trivedi. System performance with user behavior graphs. *Performance Evaluation*, 11(3):155–164, 1990.
6. G. Ciardo, A. Blakemore, Jr. P. F. Chimento, J. K. Muppala, and K. S. Trivedi. Automated generation and analysis of markov reward models using stochastic reward nets. In C. Meyer and R. Plemmons, editors, *Linear Algebra, Markov Chains and Queuing Models*, volume 48, pages 145–191. Springer-Verlag, 1993.
7. T. Dahlberg and D.P. Agrawal. Task based reliability for large systems: A hierarchical modeling approach. In *Proc. of the 22nd Intl. Conference on Parallel Processing, Volume III Algorithms & Applications*, pages 284–287, Chicago, IL, August 16-20, 1993.
8. C. R. Das and J. Kim. A unified task-based dependability model for hypercube computers. *IEEE Trans. Parallel Distrib. Syst.*, 3(3):312–324, 1992.
9. G. Florin and S. Natkin. Les reseaux de petri stochastiques. *Technique et Science Informatiques*, 4(1):143–160, 1985.
10. S. Garg, Y. Huang, C. M. R. Kintala, S. Yajnik, and K. S. Trivedi. Performance and reliability evaluation of passive replication schemes in application level fault tolerance. In *Intl. Symp. on Fault-Tolerant Computing (FTCS-29)*, June 1999.
11. Y. Huang and C. M. R. Kintala. software implemented fault tolerance: Technologies and experience. In *Intl. Symposium on Fault Tolerant Computing*, pages 2–9, Toulouse, France, June 1993.
12. M. Kaaniche, K. Kanoun, and M. Martinello. A user-perceived availability evaluation of a web based travel agency. In *Intl. Conf. on Dependable Systems and Networks (DSN'03)*, pages 709–718, San Francisco, California, June, 2003.
13. K. C. Kapur and L. R. Lamberson. *Reliability in Engineering Design*. John Wiley & Sons, New York, 1977.
14. K. W. Lee. Stochastic models for random-request availability. *IEEE Transactions on Reliability*, 49(1):80–84, March 2000.
15. E. E. Lewis. *Introduction to Reliability Engineering*. John Wiley & Sons, New York, 1987.
16. M. A. Marsan, G. Conte, and G. Balbo. A class of generalized stochastic petri nets for the performance evaluation of multiprocessor systems. *ACM Transactions on Computer Systems*, 2(2):93–122, May 1984.
17. M. K. Molloy. Performance analysis using stochastic petri nets. *IEEE Trans. on Computers*, C-31(9):913–917, Sep 1982.
18. W. Xie, H. Sun, Y. Cao, and K. S. Trivedi. Modeling of user perceived webserver availability. In *Proc. of IEEE Intl. Conf. on Communications (ICC 2003)*, Anchorage, Alaska, May 11-15, 2003.

Dependable Distributed Computing Using Free Databases

Christof Fetzer[1] and Trevor Jim[2]

[1] Technische Universität Dresden, Fakultät Informatik, Dresden, Germany
`christof.fetzer@inf.tu-dresden.de`
`http://wwwse.inf.tu-dresden.de`
[2] AT&T Labs – Research, 180 Park Avenue, Florham Park, NJ 07932, USA
`trevor@research.att.com`

Abstract. Designing and programming dependable distributed applications is very difficult. Databases provide features like transactions and replication that can help in the implementation of dependable applications. There are in particular various free databases that make it economically feasible to run a database on each computer in a system. Hence, one can partition database tables across multiple hosts to harness the processing power and disks of multiple machines. We describe a system, DOSE, that simplifies partitioning tables across multiple hosts. DOSE exposes the partitions to the programmer rather than giving the illusion of a single table. Our focus is on providing a simple implementation that works for freely-available databases, on automatic tuning of the partitions for best performance, and on applying the fault tolerance mechanisms of the databases to build dependable distributed systems. We show how we use this system to implement a distributed work queue.

1 Introduction

For a few decades researchers and practitioners have been trying to harness the power of networked computers to build systems. Different architectures and mechanisms like group communication have been proposed to achieve this. One successful example is the Google model which uses software to build a reliable system from unreliable components [5]. Using components with the best price/performance one can minimize the hardware expenses dramatically. We call such a system a RASC (Reliable and Available System from COTS components).

Mining large amounts of data is one application domain where RASCs with their cheaper but typically more unreliable hardware and software are deployed. Up to a few years ago, most large data mining jobs were running on expensive SMP machines. With the advent of cheap commodity computers, many of these jobs are now running on computing clusters.

Coping with hardware and software failures is nontrivial. The dependability community has been investigating many mechanisms like group communication, check pointing, and logging to build dependable systems to cope with failures.

M. Malek, E. Nette, and N. Suri (Eds.): ISAS 2005, LNCS 3694, pp. 123–136, 2005.

Many of these concepts are unfamiliar to application programmers who would build a RASC. Many application programmers, however, know how to write programs using databases. Using databases as the components visible to application programmers might therefore be a good match with their skills.

In this paper we are mainly interested in applications like large data mining jobs that store and query very large amounts of data in a database. To keep the promise of substantial cost reductions of RASCs, the use of free operating systems and free databases like MySQL is required. While there are commercial databases that support distributed and parallel operations, the currently available free databases are neither distributed nor parallel nor do they support distributed transactions. Furthermore, the free databases are not always very robust either. During our tests we experienced multiple database crashes and at least one of them resulted in an unrecoverable data corruption.

In this paper we introduce a middleware system (called **DOSE**: Database Oriented Systems Engineering) that supports building RASCs with the help of free databases. The idea is to run database servers on most computers in a system to harness the parallelism of the multiple CPUs, disks, and network adapters not only to maximize the throughput but also the dependability of the system.

DOSE simplifies the partitioning and replication of database tables across multiple hosts. It supports heterogeneous systems and in particular, applications using multiple computing clusters. DOSE provides a distributed request queue to simplify the implementation of worker parallelism without the need of having a centralized work dispatcher. We show how to keep the databases consistent in case of database and computer crashes using the MySQL replication mechanism that provides weak semantics.

In Section 2 we introduce the concepts underlying the design of DOSE and its API. Section 3 describes our implementation of a distributed work queue and Section 5 describes several performance measurements of DOSE. Before we conclude the paper in Section 7 we discuss related work in Section 6.

2   DOSE Concepts and API

Partitioning. Application programs can store and query data in one or more database tables. Each table consists of a finite number of rows. DOSE supports the partitioning and replication of tables. The partitioning of tables is done with the help of a *root table* (see Figure 1). The root table itself can be replicated but it is the only table that cannot be partitioned.

The percentage of rows stored in a partition is given by the weight of the partition over the sum of all weights of all partitions of the table. How the rows of a table are partitioned is under the control of the user. The idea is that a user of DOSE can define how to compute a unique MD5 hash (see RFC1321) for a row of a table. This MD5 hash is used to assign rows to partitions according to the weight of the partitions; by using the hash we get a uniform distribution for the rows. We often use the MD5 hash as the primary key but it does not need

Table	Host	Weight
Request	flagstaff	25
Request	scottsdale	25
Request	tucson	50
HostStatus	xserve-issr01	50
...	..	..

Fig. 1. The root table maps table partitions to hosts. In this example, table `Request` is partitioned across three hosts with host `tucson` getting 50% of the rows and hosts `scottsdale` and `flagstaff` each getting 25%.

to be a key nor does it need to be stored in the table. More details about the partitioning are described below.

Currently, the weights of the partitions are static during one execution. Our applications are all periodic in the sense that a brand new execution is started after the previous one has terminated. For example, a web spider application might be rerun as soon as the previous run terminates. A repartitioning can therefore be done between such executions.

Replication. MySQL supports master/slave replication. A master keeps a log file of all transactions it executes. This log file can be read and replayed by slave databases to keep their databases in sync with the master database. The fail-over from the master to one of its slaves needs additional software. We prefer solutions where the fail-over is transparent to the clients of the database. MySQL clients reconnect automatically in case the connection to the server is lost. This permits one to use a simple IP take-over mechanism.

We have previously designed and implemented a general IP take-over mechanism [4]. In case the master crashes an election determines a new master amongst the slaves of the master. The new master takes over the IP address used by the clients to connect to the database. Alternatively, there exist open source projects like HA-MySQL that provide IP take-over functionality targeted towards MySQL.

The advantage of this type of replication mechanism is the decoupling of the master and the slaves, e.g., slow slaves do not slow down the master. The trade-off is that when failing over from the master to a slave, some already committed transactions might get lost. In particular, in the context of a partitioned table this can lead to data inconsistencies, since transactions depending on such committed but lost transactions are not necessarily rolled back. We show in Section 3 how to deal with this issue in the context of the distributed work queue.

API. The DOSE API functions are written in C for increased speed (over our original PERL implementation). Most of our application code is written in PERL but the interfacing to the C code is straightforward. Each partition is represented by a handle which is needed to access the partition. We provide an iterator to iterate over all partitions of a table. However, for queries regarding a row for which the MD5 key is already known, one does not have to iterate over all partitions - instead one computes the handle of the partition via the MD5 key.

A user has control over how rows are spread across the partitions of a table. DOSE provides a function `create_md5_key` to compute an MD5 key for a row; it also returns the handle for the partition where the key is stored. The key deterministically selects a partition based on the weight of the individual partitions. To do so, we compute a mapping from the least significant byte of a key onto the set of handles of the table. For systems with large numbers of partitions one would use more than one byte.

For example, an MD5 key for a table "person" that is computed from the concatenation of two strings `familyname`, `firstname`, and some integer `ins` can be computed like this:

```
create_md5_key("person", &handle, key, "%s%s%d",
               familyname, firstname, ins);
```

The format string "%s%s%d" has the same meaning as for the C-library function `printf`; this enables **gcc** to check the type safety of `create_md5_key`'s arguments. Note that it is up to the user to select a format string and arguments that will produce a unique hash for each row. For example, if ("myers, "al", 3) and ("myer, "sal", 3) are both possible (familyname, firstname, ins) values, then the code above would give them the same hash, so, a different format string should be used. The computed key does not need to be stored in the table. However, the user can store the key in the table and use it as the primary key if this would shorten the length of the primary key.

As long as one has sufficient information to compute the MD5 key at the time of a query, one can access a row without iterating over partitions. In particular, when inserting a new row in a table there is always sufficient information to compute the MD5 key. Typically, we make sure that for all queries that use the primary key there is sufficient information to compute the partition.

Queries are written in SQL and are executed by calling the DOSE function `do_query`. For example, to retrieve all the weights for a given host from the root table, one would issue the following command:

```
do_query(handle, "SELECT weight FROM ROOT WHERE Host='%s'",host);
```

Note that the handle for the root table is returned by the DOSE initialization function.

The API function `fetchrow` retrieves the next row that was affected by the previous query. To retrieve the weight of the next row, one can write:

```
fetchrow(handle, "%d", &weight);
```

This returns 1 if the fetch is successful, and 0 if there are no more rows.

As for function `create_md5_key`, **gcc** can check the type safety of functions `do_query` and `fetchrow`.

3   Distributed Request Queue

We used the DOSE functions introduced in the previous section to implement a distributed request queue. The aim of the distributed request queue is to help distribute requests amongst a set of workers.

3.1   Concepts

The distributed request queue contains a set of requests to be processed by the workers. A worker processing a request can insert a new request into the request queue. An idle worker tries to retrieve a request from the request queue for processing. Requests can have either *at least once semantics* or *at most once semantics*. Requests with at least once semantics can and need to be reprocessed in case the worker executing the request fails. Hence, they are a little more complicated to implement than requests with at most once semantics. So far, all the requests we have implemented have had at least once semantics.

Internally, each request is associated with exactly one of the following five states (see Figure 2): *unprocessed, pending, retryable, processed,* or *failed.* A request is inserted in state *unprocessed.* When a worker retrieves the request, it becomes *pending.* When the worker successfully processes the request, the request transitions to state *processed.* If the processing of the request fails but the worker does not crash during the processing, its state becomes *failed.* A request with at least once semantics can be reexecuted after a timeout period, i.e., the state of the request transits to state *retryable.* If the processing time of a pending request approaches the timeout period, the worker executing the request can extend the timeout to prevent other workers from reprocessing the request.

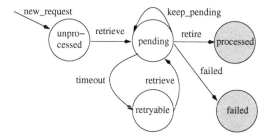

Fig. 2. States and associated state transitions for requests with at least once semantics

3.2   API

All transitions of Figure 2—with the exception of transition *timeout*—are initiated by a worker executing a corresponding API function provided by DOSE. Transition *timeout* occurs when a worker tries to retrieve a request. In this case, the retrieve function might check if a pending request has become retryable.

We provide an additional function that permits a worker to schedule the reexecution of a pending request after a given point in time. This rescheduling is mapped on the same timeout mechanism that enforces the at least once semantics of requests.

3.3   Load Manager

Each worker machine runs a load manager. The purpose of the load manager is to keep worker machines busy, but not too busy. Ideally, we would like to keep the CPUs of the worker host 100% busy. The issue is that the CPU usage of workers processing a request can change dramatically, e.g., a worker might first be blocked waiting for the transfer of some data and after it gets the data it might use all the CPU cycles it can get.

To cope with such drastically changing load conditions, a local load manager is given three load thresholds: L_{min}, L_{max} and L_{sus}. The load is the average number of processes in the running queue during the last 1 minute period. The task manager tries to keep the load above L_{min} as long as there are requests to be processed. Typically, L_{min} is chosen such that even if there are fluctuations in the CPU usage of requests, the CPUs of the worker host can still be kept 100% busy.

The load manager checks the load of the worker machine periodically (every minute). If the load drops below L_{min}, the load manager spawns new worker processes. The number of worker processes spawned is calculated based on the current estimate of the average load induced by a worker process. However, to avoid big fluctuations in the number of worker processes, we restrict the number of processes that can be spawned within a period.

When the load of a worker machine reaches L_{max}, the load manager starts signaling worker processes to terminate as soon as they finish processing their current request. The number of signaled processes is also based on the current estimate of the average load induced by a worker process and again we restrict the maximum number of worker processes that can be signaled per period.

If the load of the worker processes can change rapidly and the time to process a single request can be long, the load can rise even after the load manager started to signal worker processes that they should terminate. If the load rises above threshold L_{sus} the load manager starts suspending worker processes. As soon as the load drops below L_{sus} the load manager starts to unfreeze worker processes.

The load manager is automatically restarted when it crashes, so that it is not a single point of failure. The state of the load manager is the status of the current worker processes. Because crashes of the load manager are very infrequent, we chose a very simple recovery scheme in which all workers terminate when the load manager terminates. To achieve this, a worker process checks if the load manager has terminated before retrieving a new request. This is an efficient operation since the worker only needs to query the id of its parent process (via a system call). However, during startup the load manager must discover and unfreeze all currently suspended worker processes.

4   ACID Issues

Most databases are able to provide ACID (Atomicity, Consistency, Isolation, Durability) properties to their clients. However, these properties are not necessarily guaranteed for all likely failures (e.g., disk failure). Even if a database is replicated on a different host, a failure might cause these properties to be violated: already committed transactions can be lost (as mentioned in Section 2), i.e., the durability property can be violated. One can address some issues by adding special purpose hardware, e.g., one could use a RAID connected via a SAN to multiple hosts to mask disk failures and to be able to restart a database on a different host after a crash.

Our goal is to build a RASC, i.e., a dependable distributed computer system from cheap components. SANs and RAIDs are more expensive than adding local disks. For large systems, disk failures are non-negligible and hence, we need to address this issue in software. We describe two solutions for the distributed request queue (Section 3). First, we need to state more precisely our system and replication model.

4.1   System and Replication Model

We assume that a system contains a set of master databases. A client issues a transaction on exactly one master database. A master database has a set of backup databases. The set of backup databases is finite and might be empty. A master database logs all transaction in a replication log file and this log file is incrementally transfered to the backup databases. A backup database replays the replication log to stay in sync with the master database.

Definition 1 (received). *We say that a transaction is* received *by a backup database iff the backup database has gotten all log information from the master database that is needed to replay the transaction.*

Definition 2 (stable). *A request R is associated with a set of transactions T_R that were issued by R. We say that a request R is* stable *iff all transactions in T_R have been received by all backup databases of R.*

Definition 3 (inconsistent). *If a master database server crashes before a request is stable, switching to a backup database will make this request* inconsistent: *there exists a transaction T in T_R that was committed on the master database but that was not received on all backup databases.*

4.2   Solution

Instead of avoiding inconsistent requests, we only need to identify all inconsistent requests after a master database has crashed. Inconsistent requests with at least once semantics can be marked as unprocessed and hence, will be automatically reexecuted. If databases crashes are sufficiently rare, the expected cost of a crash can be minimized even if the execution time of a request is large. Note that inconsistent requests with at most once semantics need to be marked as failed because they cannot be reexecuted.

We introduce an approach to identify inconsistent requests. This enables us to use local replication logging with lazy flushing. The idea is that we order the retiring of requests by assigning them time stamps upon retiring and periodically determine a timestamp SR such that all requests retired before SR are stable and hence, cannot become inconsistent by a crash of a master database.

Note the assignment of a timestamp can be done very efficiently. We just store in a request R, at the time of retirement of R, the current clock value of the database that is retiring R. There is no need to synchronize the clocks of the databases or the clocks of the worker machines.

More formally, each retired request R has a timestamp $T(R)$ at which it was retired. For each partition of the request queue, we periodically compute and store a timestamp SR such that:

- for all retired requests R: $T(R) < SR \Rightarrow R$ is stable, and
- SR is maximum in the sense that if the computation of SR was initiated at time S (wrt to clock of the local database) and there exists a retired and stable request R such that $SR < T(R)$, then $S \leq T(R)$.

Operationally, we compute SR as follows. First, we insert a dummy request record into each partition of the request queue. The dummy record is assigned the local clock value of the database. Second, we determine the current log position (CLP) for each master database MS. Third, we determine the set of all slave servers for MS, and wait until all slave servers have reached position CLP. Fourth, we convert sequentially all dummy records into a "stable record marker" (without changing the timestamp) by replacing the previous "stable record marker." The value SR of a partition is given by the timestamp of the "stable record marker" stored in this partition.

Upon switch-over from a master database to a backup database each request R (with at least once semantics) that was retired at or after time SR (i.e., $T(R) \geq SR$) needs to be transitioned from state *processed* to state *unprocessed*. Rows associated with such requests can be garbage collected.

A request R with at most once semantics and $T(R) \geq SR$ needs to be transitioned to state *failed*. If this is not acceptable, a state dependent checker that verifies the completeness of a request might be a good option. This request checker can compute T_R based on the records it finds in the database that are associated with T_R. Often such a request checker is easy to implement if one knows of the need for such a checker during the design of the database tables.

Note that such an application dependent request checker is an ideal addition to the described solution: only requests with $T(R) \geq SR$ need to be check for state completeness after a crash of a master database. Only requests that are actually incomplete need to transitioned to state *unprocessed* or *failed*.

4.3   Garbage Collection

Inconsistent requests and failed requests can leave records in the databases that make the design of queries more difficult. For example, typically one wants queries to include only those rows that were written by successful requests but not by failed or inconsistent requests.

We wrote a simple garbage collector that removes all records that are not associated with a processed or pending request. When retiring a request, a key is stored within the request queue. In our applications this key is sufficient to locate all rows associated with a request. (Actually, all rows written by a request contain this key.) Our current garbage collector uses this fact but a more general garbage collector can be programmed if needed.

Our garbage collector first determines all request keys that have been stored in a row but that have not and *will not* be stored as the key of a successful request. It uses the property that requests are eventually timed-out unless a worker periodically extends the time-out of a pending request. Using the time-out mechanism the garbage collector can run in parallel with the processing of other requests and does not need to lock tables or rows.

5   Performance

We performed several performance measurements to evaluate the performance of DOSE. In our performance measurements we used a Linux cluster consisting of 8 1.8GHz Pentium 4 machines (machine names: saguaro701 – saguaro708), an SMP machine with four 1.6 GHz Xeon processors with hyperthreading (tucson), and an Apple Xserve cluster consisting of 8 machines running Mac OS X (machine names: issr01-issr08). The saguaro machines were running MySQL version 4.0.16-standard. The MAC OS X cluster and tucson were running MySQL version 4.0.16-max.

5.1   Replication Overhead

MySQL implements replication by having the master database write a log ("replication log") of all transactions that modify the database. This log is written in addition to the log file written by transaction-safe tables like InnoDB. Slaves read the replication log file via a connection to the master database. We are interested in two costs associated with replicating a database.

There is a cost associated with writing the log file and with transferring the replication log from the master to its slaves. In this measurement series we used four client machines and on each client there were 10 processes inserting requests into the request queue. There was one master database (saguaro705) and potentially one slave database (saguaro706).

We measured the throughput of five different configurations (see Figure 3): (1) without logging; (2) logging to a local file; (3) the same as (2) except running a slave database in addition, (4) the same as (2) but storing the replication log remotely, and (5) the same as (3) but storing the replication log remotely.

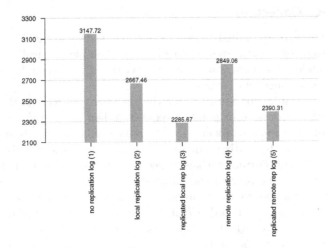

Fig. 3. Throughput of request queue for various replication configurations

Enabling the replication log resulted in a 13% throughput reduction (local disk) and about 9.5% for remote logging. Even though the remote logging adds additional network delays, the use of the additional remote disk more than compensates for the network delays. Enabling local logging and one slave reduces the throughput by about 27%. Remote logging is slightly better with a reduction of only 24%. Note that even when the log file is kept remotely, slaves read the log file through the database master.

The replication log is written before a transaction can be committed. It is our understanding (from studying the manual and the source code) that the replication log is however not flushed to disk before a transaction is committed. A synchronous flush would increase the performance penalty even further.

5.2   Partitioned Request Queue

We have investigated the performance implications of partitioning the request queue across multiple database hosts (see Figure 4). In this measurement we varied the number of partitions between 1 and 8 and the number of client machines between 1 and 6 machines. Each client machine executed 10 processes that inserted requests into the request queue. The machines used in this experiment had similar performance, and hence, we used the same weight for all partitions.

For a higher number of partitions, one client machine cannot saturate the request queue. The throughput starts to level off and falls slightly for an increasing number of partitions. If the number of client machines is however sufficiently high, the speed up of the request queue is basically linear with the number of partitions.

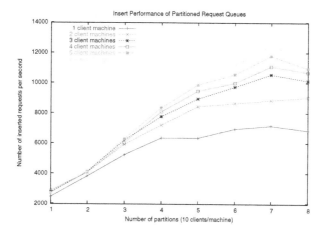

Fig. 4. Insert request queue throughput as a function of the number of partitions. The number of client machines is varied between 1 and 6 machines. Each client machine runs 10 client processes.

5.3   Load Manager

To make the job of the load manager difficult, we programmed a workload that exhibits major CPU fluctuations (see Figure 5). Each request either sleeps for up to 30 seconds or it keeps the CPU 100% busy for up to 30 seconds. We use the following load thresholds: $L_{min} = 4$, $L_{max} = 6$, $L_{sus} = 8$. The load manager keeps the load below the suspension threshold without ever actually suspending any worker.

To visualize the behavior of the load manager when it has to suspend requests, we used the same workload as in the previous measurement but we tightened the load thresholds (see Figure 6): $L_{min} = 4$, $L_{max} = 5$, $L_{sus} = 6$. The load manager had at some point to suspend two workers because it was not able to terminate workers sufficiently fast to keep the load under L_{sus}. The load manager starts to continue suspended workers as soon as the load drops below L_{sus}.

6   Related Work

Linda [3] provides a distributed shared memory called the "tuplespace" that is similar to the networked MySQL databases we use in DOSE. A tuple corresponds to a row of a table in DOSE. Linda programs can insert tuples into tuplespace and extract tuples that match a simple pattern; this is sufficient for programming things like distributed work queues. Support for tuplespace transactions and fault tolerance is built in to some variants of Linda [2,1,10,7]. DOSE uses the ACID properties of free databases like MySQL to achieve fault tolerance, and it provides a much richer query interface (SQL) than Linda's simple tuple extraction operators. DOSE's SQL interface is powerful enough for exploring the state of a distributed application in progress; for example, you can formulate an SQL query to see how many requests are *unprocessed*, or how many requests have been *processed* by a particular host. Furthermore, by exposing SQL in the DOSE API we allow the programmer to optimize the performance of tuplespace operations, as we did, for example, in our use of buckets as a way to quickly select a request from the distributed request queue described in Section 3.

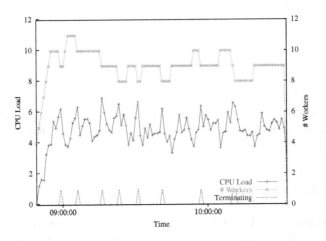

Fig. 5. Behavior of load manager for requests with random load fluctuations. ($L_{min} = 4$, $L_{max} = 6$, $L_{sus} = 8$).

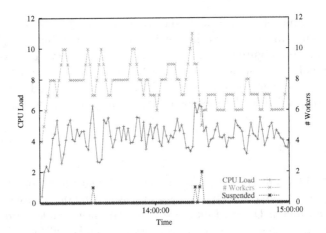

Fig. 6. When the load manager cannot keep the load under L_{sus} it starts suspending workers. It starts to continue workers as soon as the load drops below L_{sus}.

Astrolabe [8,9] is a scaleable, fault-tolerant distributed information system based on a peer-to-peer gossip protocol. The system is organized as a tree of "zones," each of which provides an SQL interface to a database of zone attributes. The database of a parent zone is defined by SQL aggregate functions over the databases of its children; thus, information about hosts (at leaf zones) is summarized towards the root. Updates propagate rapidly in the hierarchy (in tens of seconds), but Astrolabe does not support transactions as a primitive. The zone tree represents a summary of the recent state of the hosts and resources in the network, and can be used by loosely coupled applications to coordinate distributed tasks. Although both DOSE and Astrolabe use SQL, their focus is different. Astrolabe's focus is its scaleable hierarchy, which leads to its use of aggregate functions rather than arbitrary SQL to define the databases of its internal zones; the leaf zones are where Astrolabe would deploy any full-fledged SQL databases. DOSE does not have any equivalent to Astrolabe's hierarchical zones, and instead focuses on collections of hosts running commodity databases.

Commercial databases such as Microsoft SQL Server and Oracle 9i provide support for partitioning database tables across multiple servers. The different partitions are considered parts of the same virtual database table, and users can form SQL queries against the virtual table that the databases will answer by transparently sending sub-queries to each distributed partition and constructing the answer to the query from the answers to the sub-queries. In the database research community this has been studied as the problem of "answering queries using views" [6]. DOSE exposes the partitions to the programmer rather than giving the illusion of a single table; our focus is on providing a simple implementation that works for freely-available databases, on automatic tuning of the partitions for best performance, and on applying the fault tolerance mechanisms of the databases to build dependable distributed systems.

7   Conclusion

Databases are becoming commodities: they are very widely deployed, and even the cheap ones are mature and capable. Infrastructures for dependable distributed computing, on the other hand, are not commodities. Consequently, databases are benefiting from orders of magnitude more development and testing than any system for building dependable distributed applications.

DOSE is a small (currently, less than 2900 lines of code) and lightweight framework that leverages the investment being put into commodity databases towards building efficient and dependable distributed systems. Instead of building features such as transactions and replication into DOSE itself, DOSE relies on the database for them. If the commodity database does not provide sufficiently strong semantics for dependability (e.g., MySQL's imperfect replication implementation), DOSE provides just enough additional support to achieve the desired semantics. Since databases are cheap, we can deploy them on every available computer; DOSE then provides enough infrastructure (a distributed request queue,a load manager and an auto-placement of partitions) to enable programmers to easily create applications that use the available cycles efficiently.

References

1. B. Anderson and D. Shasha. Persistent Linda: Linda + transactions + query processing, 1991.
2. David E. Bakken and Richard D. Schlichting. Supporting fault-tolerant parallel programming in Linda. *IEEE Transactions on Parallel and Distributed Systems*, 6(3):287–302, 1995.
3. Nicholas Carriero and David Gelernter. *How to Write Parallel Programs: A First Course*. MIT Press, 1990.
4. Christof Fetzer and Neeraj Suri. Practical aspects of IP take-over mechanisms. In *Proceedings of 9th IEEE International Workshop on Object-oriented Real-time Dependable Systems (WORDS 2003)*, Capri Island, Italy, Oct 2003.
5. Sanjay Ghemawat, Howard Gobioff, and Shun-Tak Leung. The Google file system. In *Proceedings of the Nineteenth ACM Symposium on Operating Systems Principles*, pages 96–108, Bolton Landing, NY, October 2003. ACM Press.
6. Alon Y. Halevy. Answering queries using views: A survey. *The VLDB Journal: The International Journal on Very Large Data Bases*, 10(4):270–294, December 2001.
7. Sun Microsystems. Javaspaces service specification, version 1.1. http://wwws.sun.com/software/jini/specs/jini1.1html/js-title.html, October 2000.
8. Robbert Van Renesse, Kenneth P. Birman, and Werner Vogels. Astrolabe: A robust and scalable technology for distributed system monitoring, management, and data mining. *ACM Trans. Comput. Syst.*, 21(2):164–206, 2003.
9. Werner Vogels, Robbert van Renesse, and Ken Birman. The power of epidemics: robust communication for large-scale distributed systems. *SIGCOMM Comput. Commun. Rev.*, 33(1):131–135, 2003.
10. P. Wyckoff, S. McLaughry, T. Lehman, and D. Ford. Tspaces. *IBM Systems Journal*, 37(3):454–474, 1998.

A Compositional Framework for Real-Time Embedded Systems[*]

Insik Shin and Insup Lee

Department of Computer and Information Science,
University of Pennsylvania, Philadelphia PA 19104, USA
{ishin, lee}@cis.upenn.edu

Abstract. While component technology has been widely accepted as a methodology for designing complex systems, there are few component technologies that have been developed to accommodate the characteristics of embedded systems. Embedded systems are often subject to resource constraints as well as timing constraints. Typical scarce resources include memory for cost-sensitive systems. Many techniques, developed for reducing code size, often yield code size vs. execution time tradeoffs. Our goal is to develop a framework for supporting the compositionality of resource and timing properties. The proposed framework allows component-level resource and timing properties, which include the resource/time tradeoffs, to be independently analyzed, abstracted, and composed into the system-level resource and timing properties. In this paper, we focus on the problem of composing the collective task-level code size vs. execution time tradeoffs into a component-level code size vs. execution time tradeoff.

1 Introduction

An embedded system consists of a collection of components that interact with each other and with their environment through sensors and actuators. Current embedded systems control a range of devices, from household appliances such as refrigerators and ranges, to telephone and cellular connections, to anti-lock brakes on automobiles and cockpit displays on aircraft. In spite of growing research attention on embedded systems, there seem to be several issues that have not been adequately addressed with regard to embedded system development. In particular, even though component technology has been widely accepted as a methodology for designing complex systems, there are few component technologies that have been developed to accommodate the characteristics of embedded systems.

As embedded systems become more complex due to increased functionalities, it is necessary to develop techniques and methods that facilitate the design of large complex systems from subsystems. Component-based design has been

---

[*] This research was supported in part by NSF CCF-0429948, NSF CCR-0209024, and ARO DAAD19-01-1-0473.

M. Malek, E. Nette, and N. Suri (Eds.): ISAS 2005, LNCS 3694, pp. 137–148, 2005.

widely accepted as a methodology for designing large complex systems through systematic abstractions and compositions. Component-based design provides a means for decomposing a system into components, allowing the reduction of a single complex design problem into multiple simpler problems, and for composing components into a system through component interfaces that abstract and hide their internal complexity. Component-based design also facilitates the reuse of components that may have been developed in different environments. Current software component technologies focus on abstracting the functional aspects of components and on validating the functional aspects of assemblies of components through interfaces prior to their actual composition. However, prevailing software component technologies do not support abstraction and composition techniques for the non-functional aspects of components, e.g., timeliness, performance, reliability, safety, and resource consumption, which are critical to embedded software. Our goal is to develop a framework that supports compositionality of the non-functional aspects of components, i.e., to develop a framework where the system-level non-functional properties are established by composing together independently analyzed component-level non-functional properties.

Most embedded systems involve real-time computations. Examples include communication systems, sensor and actuator interfaces, audio and speech processing subsystems, and video subsystems. A challenging problem for real-time embedded systems is to analyze their schedulability, i.e., to determine whether the timing constraints of a real-time embedded system are all satisfiable. A key problem in designing real-time embedded systems is that most schedulability analysis techniques are not compositional. In particular, there have been no component interfaces that abstract the timing constraints of components, and thus there has been no systematic mechanism to analyze the schedulability of component assemblies through their interfaces.

Most embedded systems, unlike traditional general-purpose systems, are typically highly resource-constrained. They are often supposed to operate with limited amounts of resources, which include processor speed, memory, power, and communication bandwidth. The resource-constrained aspect of embedded systems raises issues of abstracting the resource consumption of components and predicting the resource consumption of the assembly of components before their actual composition. In this paper, we propose a composition framework where the system-level timing and resource-usage properties can be established by composing independently analyzed component-level properties.

Memory is one of the key factors that determine the manufacturing cost of an embedded system, especially when the embedded system is implemented as an SOC (System On a Chip). Many techniques have been proposed for memory size reduction by reducing program code size. One promising technique for code size reduction at the instruction set architecture (ISA) level is to use a subset of normal 32-bit instructions compressed into a 16-bit format as in ARM Thumb [8] and MIPS16 [22]. These 16-bit instructions are dynamically decompressed by hardware into 32-bit equivalent ones before execution. This approach can substantially reduce the program code size; however, it increases the number of instruction cycles needed to execute, and thus, increases the execution time of the program. For typical examples, the compressed code may require around 70% of the space of the original code, while executing 40% more instruction cycles [7].

In this paper, we introduce a compositional framework for the code size vs. execution time tradeoff, which can be obtained from the use of reduced bit-width ISA [9]. We assume that each task in a component has its own code size vs. execution time tradeoff. We present component techniques that can combine the collective task-level tradeoff information as a component-level code size vs. execution time tradeoff.

The rest of this paper is organized as follows: Section 2 describes the overview of the proposed framework and gives the system model and assumptions. Section 3 presents the problem statement, and Section 4 provides related work. Finally, we conclude in Section 5 with discussion on future research.

2 Our Compositional Framework Based on Periodic Resource Model

Our goal is to develop a compositional framework for real-time embedded systems. In this section, we present an overview of our compositional framework and provide the system models and assumptions of the framework.

2.1 Compositional Framework Overview

We consider a *compositional framework*, where components form a hierarchy and resources are allocated from a parent component to its child components in the hierarchy. The resources allocated to a single component are shared by the workloads within the component and possibly by its child components, according to a scheduling algorithm. We define a component Q as a triple (W, R, A), where W describes a set of workloads (tasks) supported in the component, R describes a set of resources available to the component, and A is a scheduling algorithm which describes how the workloads share the resources at all times.

A resource is said to be *dedicated* if it is exclusively available to a single scheduling component, or *shared* otherwise. In the proposed framework, we consider two shared resource types: a time-shared resource and a non-time-shared resource. A resource is said to be *time-shared* if it is available to a component at some times but not available at all at the other times, or *non-time-shared* if it is constantly available all the time at its partial capacity. The processor is a good example of time-shared resources, and the memory space is a good example of non-time-shared resources.

We consider that the resource requirements of a component is satisfied if the resource demands of the component is no greater than the resource supplies provided to the component. We now describe this notion more precisely. Suppose that one or more resources, R_k, $k \geq 1$, are available to a component $Q(W, R = \{R_k\}, A)$. For a component Q, its *resource demand* of a resource model R_k represents the collective resource requirements that its workload set W requests under its scheduling algorithm A. The *demand bound function* $\mathtt{dbf}_{R_k}(W, i, A, t)$ of a component Q calculates the maximum possible resource demands that W requests to satisfy the resource (timing) requirements of task i under A within a time interval of length t. The *resource supply* of a resource model R_k represents the amount of resource allocations that R_k provides. The *supply bound function* $\mathtt{sbf}_{R_k}(t)$ of R_k calculates the minimum possible resource supplies that R_k provides during a time interval of length t. A resource model R_k is said to *satisfy* a resource demand of W under A if $\mathtt{dbf}_{R_k}(W, i, A, t) \leq \mathtt{sbf}_{R_k}(t)$ for all task $i \in W$ and for all interval length t. We now define the schedulability of a scheduling component as follows: a scheduling component $Q(W, R, A)$ is said to be *schedulable*, if the minimum resource supply of R can satisfy the maximum resource demand of W under A, i.e.,

$$\forall R_k \in R \quad \forall i \in W \; \forall t > 0 \quad \mathtt{dbf}_{R_k}(W, i, A, t) \leq \mathtt{sbf}_{R_k}(t). \tag{1}$$

We define a *component abstraction* problem as the problem of abstracting the collective resource requirements, which a set of workloads demands under a scheduling algorithm, as a single resource requirement, called *resource interface*, without revealing the internal structure of the component, e.g., the number of tasks and its scheduling algorithm. We formulate the problem as follows: given a workload set W and a scheduling algorithm A, the problem is to find an "optimal" resource model R such that a component $Q(W, R, A)$ is schedulable. Here, the solution R becomes the resource interface of the component C. The optimality over the solution R can be determined with respect to various criteria such as minimizing resource capacity requirements of various resources.

In a hierarchy of components, a parent component provides resource allocations to its child components. Once a child component Q_1 finds an interface R_1, it exports the interface to its parent component. The parent component treats the resource interface R_1 as a single workload model T_1. As long as the parent component satisfies the resource requirements imposed by the single workload model T_1, the parent component is able to satisfy the resource demand of a child component Q_1. This scheme makes it possible for a parent component to supply resources to its child components without controlling (or even knowing) how the child components schedule resources internally for their own tasks.

2.2   System Models and Assumptions

As a workload model in the proposed framework, we define a task model by characterizing its requirements on two resources: the processor and the memory space. We consider that a task has a periodic real-time requirements on the processor usage and has a memory space requirement for its code size. We define a task model as $T_i \langle (P_i, C_i), S_i \rangle$ as follows:

- *Period P_i*: the fixed time interval between the arrival times of two consecutive request of T_i. We assume each task has a relative deadline equal to its period.
- *WCET (Worst-Case Execution Time) C_i*: the time amount required to complete T_i's execution in the worst case.
- *Code Size S_i*: the size of T_i's executable code.

We assume that all tasks in a component are synchronous, i.e., they release their initial jobs at the same time. We also assume that each task is independent and preemptive.

We assume that each task has multiple versions of its executable code and that each version yields different WCET and code size. To capture this, we define a *size/time tradeoff list* X_i of task T_i as follows:

- *Size/Time Tradeoff List X_i:* the list that enumerates the possible pairs of WCET C_i and code size S_i of T_i, i.e.,

$$X_i = \{x_{i,j} = \langle C_{i,j}, S_{i,j} \rangle | j = 1, 2, \ldots, K_i\} \text{ and } \langle C_i, S_i \rangle \in X_i,$$

where $x_{i,j}$ denotes the j-th element of X_i, called a size/time tradeoff, $C_{i,j}$ and $S_{i,j}$ give the WCET and code size of the j-th version of T_i's executable code, and K_i denotes the number of elements of X_i.

We assume that each task's size/time tradeoff list is derived by the selective code transformation technique [11], which utilizes a dual instruction set processor. The greedy nature of this technique ensures that the size/time tradeoff list X_i is constructed satisfying the following two properties:

- The code size $S_{i,j}$ increases while the WCET $C_{i,j}$ decreases as the index j increases. That is, $\Delta_{i,j}^S = S_{i,j} - S_{i,j-1} > 0$ and $\Delta_{i,j}^C = C_{i,j} - C_{i,j-1} < 0$, $\forall i \in [1, n]$ and $\forall j \in [2, K_i]$.
- The marginal gain in the WCET reduction for the unit increase in the code size is monotonically non-increasing, i.e.,

$$\forall i \in [1, n], \forall j \in [2, K_i - 1] \quad \frac{|\Delta_{i,j+1}^C|}{\Delta_{i,j+1}^S} \leq \frac{|\Delta_{i,j}^C|}{\Delta_{i,j}^S}.$$

Note that the minimum number of X_i's elements is 2, because $x_{i,1}$ corresponds to the program compiled entirely into the reduced instruction set, while x_{i,K_i} to the program compiled entirely into the full instruction set.

As a scheduling algorithm, we assume that the workloads within a component are scheduled under the EDF (Earliest Deadline First) scheduling policy, which has been shown to be optimal in the context of dynamic priority scheduling [15].

For a periodic task set W under EDF scheduling, Baruah et al. [4] introduced a *processor demand bound function* $\mathsf{dbf_{CPU}}(W, \mathsf{EDF}, t)$ that computes the total processor demand of W for every interval length t:

$$\mathsf{dbf_{CPU}}(W, \mathsf{EDF}, t) = \sum_{T_i \in W} \left(\left\lfloor \frac{t - D_i}{p_i} \right\rfloor + 1 \right) \cdot e_i. \tag{2}$$

For a task set W under EDF scheduling, we simply define a *memory demand bound function* $\mathsf{dbf_{SZ}}(W, \mathsf{EDF}, t)$ that computes the total memory demand of W for every interval length t:

$$\mathsf{dbf_{SZ}}(W, \mathsf{EDF}, t) = \sum_{T_i \in W} S_i. \tag{3}$$

As a resource model, we consider two resource models: one is a periodic resource model for a time-shared processor and the other is a non-time-shared resource model for the memory space. In our earlier work [19], we introduced a periodic resource model that can characterize time-shard resources with periodic behavior. This periodic resource model is defined as $\Gamma(\Pi, \Theta)$, where Π is a period ($\Pi > 0$) and Θ is a periodic allocation time ($0 < \Theta \leq \Pi$). A resource capacity U_Γ of a periodic resource $\Gamma(\Pi, \Theta)$ is Θ/Π. Let us define the supply function $\text{supply}_{\text{CPU}}(R, t, t+d)$ of a resource model R such that it calculates the processor (resource) supply of R during a time interval $[t, t + d)$. This periodic resource model $\Gamma(\Pi, \Theta)$ can specify the resources that has the following property:

$$\text{supply}_{\text{CPU}}(\Gamma, k\Pi, (k+1)\Pi) = \Theta, \quad \text{where } k = 0, 1, 2, \ldots.$$

For a periodic resource model $\Gamma(\Pi, \Theta)$, its supply bound function $\text{sbf}_\Gamma(t)$ is defined to compute the minimum resource supply for every interval length t as follows:

$$\text{sbf}_{\text{CPU}}(\Gamma, t) = \begin{cases} t - (k+1)(\Pi - \Theta) & \text{if } t \in [(k+1)\Pi - 2\Theta, \\ & \qquad (k+1)\Pi - \Theta], \\ (k-1)\Theta & \text{otherwise}, \end{cases} \qquad (4)$$

where $k = \max\left(\lceil (t - (\Pi - \Theta))/\Pi \rceil, 1\right)$.

We define a non-time-shared memory resource model $\Psi(\Upsilon)$, where Υ represents the memory size that the memory model Ψ can provide. For this model $\Psi(\Upsilon)$, its supply bound function $\text{sbf}_\Psi(t)$ is simply defined as

$$\text{sbf}_{\text{SZ}}(\Psi, t) = \Upsilon.$$

We define an interface model as $\mathcal{I}\langle(\mathcal{P}, \mathcal{C}), \mathcal{S}\rangle$, where $\mathcal{P}$ is a period, $\mathcal{C}$ is a WCET, and $\mathcal{S}$ is a code size.

Definition 1. *An interface* $\mathcal{I}\langle(\mathcal{P}, \mathcal{C}), \mathcal{S}\rangle$ *is said to* abstract *the resource demands of a component* $Q(W, R, A)$, *denoted by* $\mathcal{I} \models Q$, *if*

$$\forall t > 0 \qquad \text{dbf}_{\text{CPU}}(W, \text{EDF}, t) \leq \text{sbf}_{\text{CPU}}(\Gamma(\Pi, \Theta), t) \quad \wedge$$
$$\text{dbf}_{\text{SZ}}(W, \text{EDF}, t) \leq \text{sbf}_{\text{SZ}}(\Psi(\Upsilon), t),$$

where $\Pi = \mathcal{P}, \Theta = \mathcal{C}$, *and* $\Upsilon = \mathcal{S}$.

We consider each task T_i has multiple candidates of a pair of WCET C_i and code size S_i, and we defined a size/time tradeoff list $X_i = \{\langle C_{i,j}, S_{i,j}\rangle\}$ to specify these multiple candidates. Like an individual task, an interface $\mathcal{I}$ can have multiple candidates of a pair of WCET $\mathcal{C}$ and code size $\mathcal{S}$ as well. To capture this, we define a size/time tradeoff list $\mathcal{X}$ of $\mathcal{I}$ such that $\mathcal{X}$ enumerates possible pairs of $\mathcal{C}$ and $\mathcal{S}$, i.e., .

$$\mathcal{X} = \{\langle \mathcal{C}_m, \mathcal{S}_m\rangle | m = 1, 2, \ldots, \mathcal{K}\} \text{ and } \langle \mathcal{C}, \mathcal{S}\rangle \in \mathcal{X},$$

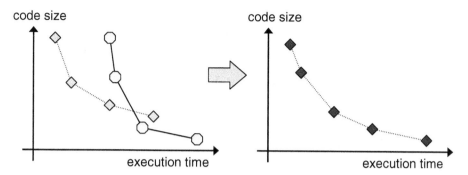

Fig. 1. Component abstract problem with size and time

where $\mathcal{C}_m$ and $\mathcal{S}_m$ give the m-th WCET and code size, and $\mathcal{K}$ denotes the number of elements of $\mathcal{X}$.

We classify the elements of $\mathcal{X}$ into three categories. An element $\langle \mathcal{C}, \mathcal{S} \rangle$ of $\mathcal{X}$ is said to be *dominated* if

$$\exists \langle \mathcal{C}', \mathcal{S}' \rangle \in X \quad \mathcal{C}' \leq \mathcal{C} \text{ and } \mathcal{S}' \leq \mathcal{S}.$$

An element $\langle \mathcal{C}_m, \mathcal{S}_m \rangle$ of $\mathcal{X}$ is said to be *convex*, where $1 < m < \mathcal{K}$, if

$$\forall i \in [1, m-1] \ \forall j \in [m+1, \mathcal{K}] \ \frac{|\mathcal{C}_m - \mathcal{C}_i|}{\mathcal{S}_m - \mathcal{S}_i} \leq \frac{|\mathcal{C}_j - \mathcal{C}_m|}{\mathcal{S}_j - \mathcal{S}_m}.$$

The first element $\langle \mathcal{C}_1, \mathcal{S}_1 \rangle$ and the last element $\langle \mathcal{C}_\mathcal{K}, \mathcal{S}_\mathcal{K} \rangle$ are convex if they are not dominated, respectively. An element $\langle \mathcal{C}_m, \mathcal{S}_m \rangle$ of $\mathcal{X}$ is said to be *inbetween* if it is neither dominated nor convex.

3 Extension to Component Abstraction on Code Size and Timeliness

In this paper, we present an extension to our compositional framework for components with size and time. With the system model and assumptions described in the previous section, we first present the formal statement of our problem, called the CAP-ST (component abstraction problem: size and time) problem.

Consider a component $Q(W, R, A)$, where $W = \{T_i(\langle P_i, C_i \rangle, S_i)\}$, $i = 1$, $\ldots, n$, $A = \mathsf{EDF}$. Basically, we consider the problem of abstracting the resource information of Q with an interface $\mathcal{I}(\langle \mathcal{P}, \mathcal{C} \rangle, \mathcal{S})$. We assume that each task T_i has a size/time tradeoff list $X_i = \{\langle C_i, S_i \rangle\}$ to represent its own size and time tradeoff information. Here, we are particularly interested in abstracting the collective task-level size and time tradeoff information as a component-level size and time tradeoff information. That is, we want to compose multiple size/time tradeoff lists X_i's as a single size/time tradeoff list $\mathcal{X} = \{\langle \mathcal{C}, \mathcal{S} \rangle\}$, as illustrated in Figure 1. For this abstraction problem, we assume that the period P_i of each task T_i is given and the period $\mathcal{P}$ of an interface $\mathcal{I}$ is given. We define the CAP-ST problem as follows: given P_i and X_i for each task T_i and $\mathcal{P}$, the problem is to construct $\mathcal{X}$ such that

$$\forall T_i \in W \quad \forall \langle C_i, S_i \rangle \in X_i \quad \exists \langle \mathcal{C}, \mathcal{S} \rangle \in \mathcal{X} \qquad \mathcal{I}(\langle \mathcal{P}, \mathcal{C} \rangle, \mathcal{S}) \models Q(W, R, A).$$

We now present an approach to the CAP-ST problem with the following example. Consider a component $Q(W, R, A)$, where $W = \{T_i \mid i = 1, 2\}$ and $A = \mathsf{EDF}$. Let $p_1 = 50$ and $p_2 = 70$. Suppose the size/cycle tradeoff lists X_1 and X_2 are given such that $\langle c_i, s_i \rangle \in X_i$, where $i = 1, 2$, as follows:

$$X_1 = \{\langle 3.47, 0.64 \rangle, \langle 3.04, 0.69 \rangle, \langle 2.80, 0.78 \rangle, \langle 2.69, 0.84 \rangle\},$$
$$X_2 = \{\langle 4.46, 1.55 \rangle, \langle 4.02, 1.64 \rangle, \langle 3.92, 1.71 \rangle, \langle 3.84, 1.80 \rangle\}.$$

Now, we can construct the size/time tradeoff list $\mathcal{X} = \{\langle \mathcal{C}, \mathcal{S} \rangle\}$ of the example component as follows:

- WCET $\mathcal{C}$. We find the minimum possible value of $\mathcal{C}$ that, given $\mathcal{P}$, satisfies

$$\forall 0 < t \quad \mathsf{dbf}_{\mathsf{CPU}}(W, \mathsf{EDF}, t) \leq \mathsf{sbf}_{\mathsf{CPU}}(\Gamma(\Pi, \Theta), t). \qquad (5)$$

  where $\Pi = \mathcal{P}$ and $\Theta = \mathcal{C}$. Suppose each task T_i has a WCET c_i determined as one of the candidate values given by X_i. For example, let $c_1 = 3.47$ and $c_2 = 4.46$. In this example, when $\mathcal{P} = 25$, the minimum value of $\mathcal{C}$ to satisfy Eq. (5) is 1.63.
- Code Size $\mathcal{S}$. We simply determine $\mathcal{S}$ as follows:

$$S = \sum_{\tau_i \in W} s_i,$$

  and this surely satisfies

$$\mathsf{dbf}_{\mathsf{SZ}}(W, \mathsf{EDF}, t) \leq \mathsf{sbf}_{\mathsf{SZ}}(\Psi(\Upsilon), t),$$

  where $\Upsilon = \mathcal{S}$. Suppose each task $T_i \in W$ has a code size s_i determined as one of the candidates given by X_i. For instance, $s_1 = 0.64$ and $s_2 = 1.55$. Then, $\mathcal{S} = 2.19$.

We now consider an issue of constructing the size/time tradeoff list $\mathcal{X}$. When each task has multiple elements of its size/time tradeoff list, a pair of $\langle \mathcal{C}, \mathcal{S} \rangle$ has multiple candidate values. For the above example with $\mathcal{P} = 25$, Figure 2 shows all the possible elements of X under the labels of "convex", "inbetween", and "dominated". The labels indicate which categories elements belong to. The size/cycle tradeoff list $\mathcal{X}$ can be refined such that dominated elements are excluded or $\mathcal{X}$ contains only convex elements. The latter way ensures that X satisfies the following property: the marginal gain in the WCET reduction for the unit increase in the code size is monotonically non-increasing, as each task-level size/cycle tradeoff list X_i does.

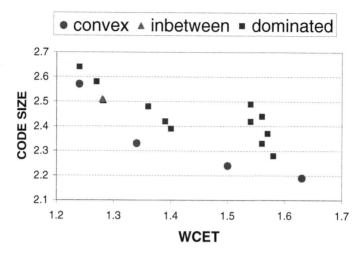

Fig. 2. The elements of the solution $\mathcal{X}$

4   Related Work

4.1   Code Size Reduction Techniques

For many embedded systems, program code size is a critical design factor. We present a brief overview of a compiler technique for code size reduction that works for a processor capable of executing dual bit-width instructions. A very good example of such a processor is ARM microprocessors with a 32-bit instruction set (IS) for normal modes and a 16-bit reduced bit-width IS for Thumb modes [8]. A reduction in code size comes from encoding a subset of the 32-bit normal mode IS into the 16-bit Thumb mode IS. At the execution time, a decompression engine converts a Thumb-mode instruction into an equivalent normal-mode instruction during the decode stage. The Thumb IS can access only 8 general purpose registers (out of 16 general purpose registers in the normal mode) and can encode only a small immediate value. These limitations increase the number of execution cycles and, thus, increases the program execution time. For typical programs, by using this technique the code size can be reduced by around 30%, while the number of execution cycles increases by about 40% [7].

The dual bit-width ISA allows a program to contain both 32-bit normal-mode instructions and 16-bit reduced bit-width instructions where the mode change between the two can be performed by executing a single mode-change instruction. This capability allows for a tradeoff between code size and the number of execution cycles when compiling a program. For example, by progressively transforming program units such as functions or basic blocks in the normal mode into the equivalent ones in the reduced bit-width mode while adding patch-up code to maintain the correct semantics, we can obtain a table that gives possible (code size, the number of execution cycles) pairs. The order by which the transformation is performed considers both reduction in code size and increase in the the number of execution cycles, i.e., it favors program units that give large reduction in code size with only a small increase in the number of execution cycles. Our earlier work [21] proposed a design framework that deals with a design problem taking advantage of this code size vs. time tradeoff. In this paper, we introduce a new abstraction technique that addresses the issues of composing the collective task-level code size vs. time tradeoff information into a component-level code size vs. time tradeoff information.

4.2   Component Techniques for Timing Aspect

Real-time systems are ones in which correctness depends not only on logical correctness but also on timeliness. In the real-time systems community, substantial research efforts have concentrated on the schedulability analysis problem, which determines whether timing requirements imposed on the system can be satisfied. For example, extensive studies [15,12,3,2] have been conducted the schedulability analysis for dedicated systems. In addition, the schedulability analysis on hierarchical scheduling frameworks, where components (applications) can share resources hierarchically under different scheduling, has been receiving a growing attention [5,10,13,6,17,18,14,19,1,20]. However, there is no widely accepted technique that supports the compositionality of timing requirements, i.e., how component-level timing requirements can be independently analyzed, abstracted, and composed into the system-level timing requirements.

We have developed a compositional real-time scheduling framework [14,19] for supporting the compositionality of timing requirements. Fundamental to such a framework is the problem of computing the minimum resource requirements necessary for guaranteeing the collective timing requirements of a component or a component assembly. We have addressed this problem systematically, by developing sufficient and necessary schedulability conditions for the two most popular real-time scheduling algorithms: EDF (earliest deadline first) and RM (rate-monotonic). We addressed this problem using a standard real-time requirement model, which is the Liu and Layland periodic model [15], and another model, the bounded-delay resource partition model [16].

5   Conclusion

Our goal is to develop a framework for supporting the compositional modeling and analysis of timing and resource consumption properties. In this paper, we considered the problem of supporting the compositionality of timing and code size properties. Particularly, we focused on the problem of composing the collective task-level code size vs. execution time tradeoff into a component-level code size vs. execution time tradeoff.

Our future work includes extending our framework by considering other resources, such as power. For example, the dynamic voltage scaling (DVS) technique which involves dynamically adjusting the supply voltage and the CPU clock speed, has been widely accepted as a key technique to reduce the energy consumption for embedded systems. This DVS technique generates energy consumption vs. execution time tradeoff. We plan to develop component techniques that support the compositionality of this energy consumption vs. execution time tradeoff.

In this paper, we assume that each task is independent. However, tasks may interact with each other through communication and synchronization. We also consider extending our framework to deal with this issue.

References

1. L. Almeida and P. Pedreiras. Scheduling within temporal partitions: response-time analysis and server design. In *Proc. of the Fourth ACM International Conference on Embedded Software*, September 2004.
2. N. Audsley, A. Burns, and A. Wellings. Deadline monotonic scheduling theory and application. *Control Engineering Practice*, 1(1):71–78, 1993.
3. S. Baruah, R. Howell, and L. Rosier. Algorithms and complexity concerning the preemptive scheduling of periodic, real-time tasks on one processor. *Journal of Real-Time Systems*, 2:301–324, 1990.
4. S. Baruah, A. Mok, and L. Rosier. Preemptively scheduling hard-real-time sporadic tasks on one processor. In *Proc. of IEEE Real-Time Systems Symposium*, pages 182–190, December 1990.
5. Z. Deng and J. W.-S. Liu. Scheduling real-time applications in an open environment. In *Proc. of IEEE Real-Time Systems Symposium*, pages 308–319, December 1997.
6. X. Feng and A. Mok. A model of hierarchical real-time virtual resources. In *Proc. of IEEE Real-Time Systems Symposium*, pages 26–35, December 2002.
7. S. Furber. *ARM System Architecture*. Addison Wisley, New York, NY, 1996.
8. L. Goudge and S. Segars. Thumb: Reducing the cost of 32-bit RISC performance in portable and consumer applications. In *Proc. of the 1996 COMPCON*, September 1996.
9. A. Halambi, A. Shrivastava, P. Biswas, N. Dutt, and A. Nicolau. An efficient compiler technique for code size reduction using reduced bit-width isas. In *Proceedings of Design Automation and Test in Europe (DATE '02)*, 2002.
10. T.-W. Kuo and C.H. Li. A fixed-priority-driven open environment for real-time applications. In *Proc. of IEEE Real-Time Systems Symposium*, pages 256–267, December 1999.
11. S. Lee, J. Lee, C. Y. Park, and S. L. Min. A flexible tradeoff between code size and WCET using a dual instruction set processor. In *Proceedings of the 8th International Workshop on Software and Compilers for Embedded Systems (SCOPES)*, pages 244–258, Amsterdam, The Netherlands, September 2004.
12. J. Lehoczky, L. Sha, and Y. Ding. The rate monotonic scheduling algorithm: exact characterization and average case behavior. In *Proc. of IEEE Real-Time Systems Symposium*, pages 166–171, 1989.
13. G. Lipari and S. Baruah. A hierarchical extension to the constant bandwidth server framework. In *Proc. of IEEE Real-Time Technology and Applications Symposium*, pages 26–35, May 2001.
14. G. Lipari and E. Bini. Resource partitioning among real-time applications. In *Proc. of Euromicro Conference on Real-Time Systems*, July 2003.

15. C.L. Liu and J.W. Layland. Scheduling algorithms for multi-programming in a hard-real-time environment. *Journal of the ACM*, 20(1):46 – 61, 1973.
16. A. Mok, X. Feng, and D. Chen. Resource partition for real-time systems. In *Proc. of IEEE Real-Time Technology and Applications Symposium*, pages 75–84, May 2001.
17. J. Regehr and J. Stankovic. HLS: A framework for composing soft real-time schedulers. In *Proc. of IEEE Real-Time Systems Symposium*, pages 3–14, December 2001.
18. S. Saewong, R. Rajkumar, J.P. Lehoczky, and M.H. Klein. Analysis of hierarchical fixed-priority scheduling. In *Proc. of Euromicro Conference on Real-Time Systems*, June 2002.
19. I. Shin and I. Lee. Periodic resource model for compositional real-time guarantees. In *Proc. of IEEE Real-Time Systems Symposium*, pages 2–13, December 2003.
20. I. Shin and I. Lee. Compositional real-time scheduling framework. In *Proc. of IEEE Real-Time Systems Symposium*, December 2004.
21. I. Shin, I. Lee, and S. Min. Embedded system design framework for minimizing code size and guaranteeing real-time requirements. In *Proc. of IEEE Real-Time Systems Symposium*, pages 201–211, December 2002.
22. D. Sweetman. *See MIPS Run*. Morgan Kaufmann, San Francisco, CA, 1999.

On the Importance of Composability of Ad Hoc Mobile Middleware and Trust Management

Ovidiu V. Drugan[1], Ioanna Dionysiou[2], David E. Bakken[1,2], Thomas P. Plagemann[1],
Carl H. Hauser[2], and Deborah A. Frincke[3]

[1] Department of Informatics, University of Oslo, Norway
{ovidiu, dbakken, plageman}@ifi.uio.no
[2] School of Electrical Engineering and Computer Science, Washington State University,
Pullman, Washington, USA
{idionysi, bakken, hauser}@eecs.wsu.edu
[3] CyberSecurity Group, Pacific Northwest National Laboratory, Richland, WA, USA
deborah.frincke@pnl.gov

Abstract. Distributed computing is widely expected to become ubiquitous over
the next decade. Distributed services such as those provided by Service Oriented
Architectures which will support this ubiquity must meet many requirements at
both runtime and over their lifecycle. Composability is one key requirement for
such services. In this paper we provide refined definitions of composability as it
applies to such distributed services, encompassing both hardware and software.
We then analyze these composability issues as they apply to two middleware
frameworks which support critical infrastructures. The first examines composability issues for resource management infrastructure for a framework that provides middleware services for ad hoc mobile networks designed to support emergency rescue coordination. The second investigates composability issues involved
with trust management for status dissemination for the electric power grid.

1 Introduction

In the last ten to fifteen years distributed computing has become mainstream. It has
transitioned from something barely more than a laboratory curiosity to being relied
on in all facets of society. This transition has been driven by many factors, including
cheap computing hardware, the availability of broadband internet connections, and the
maturing of commercial middleware.

Over the next decade distributed computing is widely predicted to become ubiquitous and the default mode of interaction for most application programs. These predictions are being driven by factors including the rapidly decreasing size and cost of
networked embedded computing hardware and the widespread availability of wireless
network connections. Distributed computing over the next decade will not only become
more widespread in homes, schools, shopping malls, and many public places, but also
in critical infrastructures. Many of these infrastructures such as the electric power grid
currently have very rudimentary communications and application-level services, but are
undergoing ambitious modernization efforts, a large part of which includes improving
their communications infrastructure, including middleware services.

M. Malek, E. Nette, and N. Suri (Eds.): ISAS 2005, LNCS 3694, pp. 149–163, 2005.

Programming distributed applications is even harder than programming standalone ones. Fundamental factors inherent in wide-area distributed computing cause this, chiefly having to deal with the variability of network-level latencies and partial failures of computing nodes and network links. At the application level, distributed applications increasingly have complex quality of service (QoS) requirements with multiple QoS dimensions such as latency, throughput, availability, confidentiality, and integrity [1].

Distributed applications have historically been expensive to develop and may be deployed for many years. Thus, developers, system integrators, and maintainers of such software have had additional higher-level requirements for their software systems. These include flexibility, adaptability, survivability, manageability, evolvability, and composability. Composability in particular is a key requirement that has become of great interest in recent years [2,3,4]. It involves reasoning about and providing end-to-end interoperability and QoS across different entities, in ways we delineate in the following section.

In this paper we investigate the space of serial composability for distributed application programs. To help clarify this and make it more concrete, we illustrate these issues in two very different contexts. The first is serial composability of resource management for ad hoc mobile environments, driven by emergency response application needs. The second context is composability of trust management, particularly serial trust composability, in the context of supporting middleware services for critical infrastructures, primarily the electric power grid.

2   Facets of Composability

There are many different ways in which composability is desired for distributed application programs. In this section we provide expanded definitions of such kinds of composability.

Hierarchical composability involves composing up from contained components. A given component has a functional API, that deals with its business logic, as well as (either explicitly or implicity) a QoS interface. Hierarchical composition of functional APIs has been practiced for decades and is known as the "divide and conquer" technique. However, when a component uses other local components, the QoS provided by these subcomponents should ideally be composed upward. That is, the QoS and resource usage of the subcomponents plus that of the component body's code should be composed into the QoS and resource usage for the higher level component. Such composability is an area of active systems software research, and is currently beyond the state of the art. It is sometimes called the "system of systems" problem.

Horizontal composability involves composition across peer entities which can be composed in two primary ways: in parallel and in sequence. *Parallel composability* involves supporting multiple entities competing for systems mechanisms and middleware services without interfering with the delivery of the end-to-end QoS that each running application client receives. Parallel composability issues and techniques are described in [2]. *Serial composability* is the ability to provide end-to-end services with predictable QoS for application clients that are composed of a chain of system mechanisms. Providing serial composability of the above entities is challenging and in the general case beyond the state of the art. There are many open research issues involving the necessary support infrastructure for this, including simulations and testbed experimentation infrastructures that have necessary hooks to validate the composability of QoS.

Vertical composability involves composing composition of different abstraction levels (i.e., not peers), often in a stack and in the same process. Examples of these levels, from the bottom up, include:

- *Baseline system mechanisms*, which are low-level mechanisms that can be considered atomic for the purposes of composability analysis. Examples include process creation, process scheduling, bandwidth reservation, and memory allocation.
- *Compound system mechanisms*, which are built on top of the baseline system mechanism. These include replication, checkpointing, and process migration.
- *Middleware*, on which application programs are built.
- *Application programs*, both clients and servers.

The middle two of these layer–compound system mechanisms and middleware–may themselves involve multiple layers across which vertical composability must be provided. [5] describes how replication using the state machine approach depends on lower-level compound system mechanisms including resilient processes, RPC, and multicast. [6] describes 4 layers of middleware; from the bottom upwards they are: host infrastructure middleware, distribution middleware, common middleware services, and domain-specific middleware.

Resource management can compose in all 3 ways described above, though typically not all in a single resource management system. Vertical composition involves composing resource usage up a stack. For example, classical resource managers may allocate low-level resources such as memory and CPU time, while a middleware manager will create and manage distributed objects (or whatever abstraction the given middleware provides) based on the lower-level resources provided by the baseline resource manager. Serial composition of resources is important in the ad hoc mobile network domain, as described in Section 3. Hierarchical resource management is very important for complex wide area distributed application programs such as for the military. An example of a hierarchical resource manager is Darwin [7].

3   Mobile Resource Management and Composability

3.1   Overview of the Ad Hoc InfoWare Project

Efficient collaboration between rescue personnel from various organizations is a mission critical key element for a successful operation in emergency and rescue situations. There are two central preconditions for efficient collaboration, (1) the incentive to collaborate, which is naturally given for rescue personnel in many emergency response situations which may involve fire, police, and paramedics; and, if terrorists are involved, collaboration may also involve military, disease control experts, and nuclear or chemical experts and (2) the ability to efficiently communicate and share information. Mobile ad hoc networks (MANETs) have the potential to provide the technical platform for efficient information sharing in such scenarios, assuming that all rescue personnel are carrying and using mobile computing devices with wireless network interfaces. Applications are needed to turn a working infrastructure of a MANET into a useful system, like dispatching of rescue personnel and equipment, context-aware medical diagnosis and treatment support, and real-time evidence collection and management. Unfortunately, the state of the art in such middleware for MANETs is very limited, with almost no support for the necessary MANET-specific interaction styles and resource management. As a result, application development for MANETs is particularly difficult.

MANETs are typically highly dynamic networks in terms of available communication partners, available network resources, connectivity, etc. Furthermore, the end-user devices are very heterogeneous, ranging from high-end laptops to low-end PDAs and mobile phones. CPU, storage space, bandwidth, and battery power represent important resources. Finally, many application scenarios, like coordination of rescue teams, have hard non-functional requirements such as availability, efficient resource utilization, security, and privacy. Both the heterogeneity of devices and the broad range of functional and non-functional requirements make a composable solution preferable. Complex middleware services are decomposed in such a way that on resource weak devices only some of the service components are deployed and on resource strong devices all service components are deployed to meet the application requirements. Further, serial composability is an issue: these non-functional requirements must often be provided across multiple infrastructures. Thus, sufficient quality in information access and sharing in such an environment faces many obstacles. Obviously, solving these issues in every new MANET application from scratch is not practical, nor would it be desirable even if it were practical. Rather, a set of composable middleware services that support the development of applications for MANETs is needed.

Our goal in the Ad Hoc InfoWare project is to develop composable middleware services for information sharing in MANETS, with a key driving example being emergency and rescue operations. We assume that wireless computing devices will be used as the basic technical means for information sharing between rescue personnel such as policemen, firemen, physicians, and paramedics.

These MANETs at emergency sites have the typical MANET complications such as heterogeneous nodes, unpredictable reachability of nodes, etc. However, in many cases MANETs at emergency sites will not be entirely infrastructureless, because some devices might serve as gateways to the Internet. Middleware services for MANETs must thus provide serial composability of QoS and resource management policies that span this wired-wireless environment.

We address these challenges and requirements in the Ad-Hoc InfoWare project by developing a set of configurable middleware components for MANETs that provide their services to applications and to other middleware components. Figure 1 illustrates our architecture, comprising five major components: *knowledge management, distributed event notification, watch dogs, resource management,* and *security and privacy management*. The knowledge management component supports sharing of information by accommodating the heterogeneity of organizations involved. It presents the information in a way that human and non-human users of all organizations can understand. This implies supporting functionality akin to high-level distributed database system functionality, querying available information and keeping track of what information is available in the network. The likelihood of connection loss, at best sporadically and sometimes quite regularly, implies also that middleware services based on synchronous communication are not a good choice, because they are too vulnerable with respect to communication disruptions. The alternative to synchronous interactions is a distributed event notification system (DENS). Events are detected by watchdogs on the devices and the DENS delivers notifications as reliably as possible to their destinations. Devices will inevitably lose contact with other mobile devices due to network partitioning or power drain, but groups that are portioned off from other parts of the MANET must still function as well as possible. Therefore, replication is necessary to achieve the required level of availability. This includes replication of data as well as replication of middleware services, e.g., DENS functionality and state. In order to make replication decisions that increase the availability and result in efficient resource utilization, it is important to keep track of resources. This resource management (RM) must track resources across different wired and wireless infrastructures and must support policies that compose serially across infrastructures to provide end-to-end QoS for the wide range of applications, users, and corresponding usage patterns.

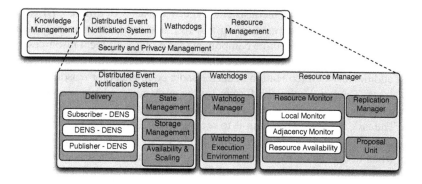

Fig. 1. Ad-Hoc InfoWare middleware components

Performance and efficient resource utilization are also important, but there is typically a trade-off between these two requirements and availability. There is no general solution for this trade-off and its resolution often depends on the particular application and even the particular emergency situation. It is thus necessary to allow the application (or system integrator or system administrator) to define policies on how to handle these tradeoffs. Hardware heterogeneity requires also that middleware services are configurable, such that small resource-weak devices run only a few middleware components and devices with sufficient resources run many (or all) components. This is a different kind of serial composability, of services across a span of device capability. Careful management of this can prolong the lifetime of the resource-weak devices and thereby provide broader coverage of the service for users of the application. The security and privacy management component is concerned with controlling access to shared information.

3.2 Resource Management

The Ad Hoc InfoWare middleware is designed as a set of configurable components to address the wide range of devices. Configuration of components is achieved by decomposing each component into a set of sub-components. Each component can deploy only a part of its sub-components or all. By this it can adapt to the available resources. Obviously, a minimal configuration, i.e., deploying only one sub-component, can only provide the basic functionality. Richer configurations that include additional sub-components can provide more functionality but requires more resources.

We illustrate the possible configuration for the Ad Hoc InfoWare middleware by describing the DENS and RM. DENS two main tasks are delivery of notification and state management. Delivery is built on three delivery components to exchange information on subscriptions and notifications between the specialized three pairs of entities: subscriber - DENS, DENS - DENS, and DENS - publisher. A minimal configuration contains only one of the delivery components, which is needed by any node that wants to use and or provide DENS related services. A richer configuration for DENS also involves the three management components, i.e., state management, storage management, and availability and scaling management [8]. These management components allow storing subscriptions and notifications that could not be delivered due to network partitions, and handling inconsistencies between states of different DENS nodes. RM main tasks are resource monitoring and resource information management. Resource monitoring is built on three components for monitoring local resources, remote resources and dissemination of resource information. A richer RM configuration could consist additionally of a replication manager and a component for reasoning about use of resources. The latter helps the former by recommending whether to use or to free resources, respectively, e.g., where to replicate data, or which resource to use on other nodes. The task of the proposal unit is to recommend to the replication manager when to replicate data and to which node. This implies also that the proposal unit recommends when to use which resources and when to free them. These management components allow applications to use remote resources and increase availability of data in the network.

In order to provide a complete and correct set of middleware services it is necessary that each network, including every partition of the network, hosts a complete version of each of the middleware services. The role of RM in such an environment is to know what services a node is running and what services are missing. It needs to find ways to balance the lack of resources on nodes, by composing the service instances running on a set of nodes. One important factor in composing the service is the amount of resources: on the local node, on another node, and on a set of nodes.

We illustrate this with four examples where service composability gives the possibility to middleware and applications to react correctly to changes in the environment and by this to provide better services to the user.

- *Composing resources for DENS - Ad Hoc InfoWare middleware*
  In case of a network partition a middleware service, such as DENS, might not have any instance. In this case, it is the task of RM to discover one or more nodes able to run the missing sub-services of DENS. If the service is missing completely it can instantiate the entire DENS service.

- *A smart resource usage by using knowledge - Ad Hoc InfoWare middleware*
  Recommending use of a resource from another node raises trust and reliability problems. To avoid such problems RM can request from the KM high level information about a node's grouping information, for example the owner organization. It selects for recommendation the node with the highest degree of trust.
- *Using video automatic analysis for evidence collection - application*
  We envision an application for online automatic annotation and analysis of video recordings. Such an application could be composed from the following three services: video recording, online video analysis and online video annotation. Each of these services can be offered by a different node. If the capture node is not able to perform all tasks, they might be distributed to other nodes. This can happen for various reasons, for example the service is not present locally or the it doesn't have enough computation power. In this case, the application can request to RM to find the nodes where these services can be performed.
- *Advanced video streaming and transcoding - application*
  Another possible application is video streaming and transcoding for clients with different capabilities. For such an application the following components can be envisioned: high resolution video recording, transcoding service, storage service and streaming service. If the video capture node is unable to perform a video transcoding and streaming, the application can request the RM to find the nodes where these services can be performed.

In these examples, although the source of the request is different, the goal of the service is the same: to find the resource that suits best its requirements, which is the task of RM. A way of achieving this is to monitor and predict data and resource availability.

To predict availability of resources the RM first predicts the future connectivity to the nodes. The second step is to disseminate the resource availability information in the network. To determine the connectivity to nodes it is useful to estimate their current and future position. A way to perform this is to determine the mobility patterns for each node and group of nodes, which means determining the pattern of movement in a scene for a node. One constraint of our application domain is that it is not possible to have exact location information on all nodes all the time, because GPS devices will not always work (e.g. in tunnels or buildings). Due to this, it can rely only on other types of information, for example routing tables from the routing protocol, wireless bandwidth characteristics, statistics of adjacency of nodes or group membership descriptions.

Currently, we are working on analyzing data obtained from the Ad hoc On-Demand Distance Vector (AODV) Routing protocol [9] for prediction of future connectivity of nodes. Each node keeps track of its neighbors, but since AODV is a reactive routing protocol, only as long as it is involved in communications, i.e., transmitting and receiving data, or transmitting routing messages. We use this information to build neighbors' histories, which we later use to estimate the future connectivity between two neighboring nodes. This method has the advantage that it can be used for any other routing protocol. Another advantage is that it doesn't create extra load on the system, since it does not send any messages but only monitors the routing tables for changes. Our preliminary experiments have shown that we can predict connectivity of nodes up to 170 seconds in the future. We also can use application level information like maps and location services if present and reachable in the network. The second step when predicting data and resource availability is to disseminate and search available resources and data. In order to disseminate information on availability of resource and data, we currently investigate the use of DENS and the direct collaboration between RM instances on different nodes.

3.3   Related Work to Resource Management for Middleware for MANETs

Most of the existing work on resource management in ad hoc networks is oriented toward studies of QoS (Quality of Service) [10,11,12], bandwidth management [13] and mobility management [14].

Some of the existing work proposes the use of nodes' mobility information to improve information accessibility in MANETs. For example, Chang et al [15] propose a framework for a distributed data accessibility service to access multimedia data within a heterogeneous cooperative group. It is assisted by a predictive location-based routing protocol which tries to maintain a specific set of QoS parameters. For this, they assume that moving nodes remain in the same groups and follow predictable movement patterns. Each node constructs the movement patterns of its neighboring nodes. For this, it relies on information like the geographic location of nodes, movement direction and velocity, transmission range of the node and on the received periodic position broadcasted from the nodes. Using movement patterns, each node participating in a transmission is capable of predicting the future location of the intermediate nodes and destination. Under similar assumptions NonStop [16] constructs the movement patterns for a set of mobile nodes which exhibit similar mobility patterns in their movements. They are used to guarantee the continuous availability of multimedia streaming. NonStop estimates the occurrence of network partitioning to replicate data to a streaming server that has a low probability of being disconnected from a requesting client during a streaming session. For optimization, MARE [17] tries to reduce bandwidth requirements by moving operations, rather than transmitting data across a network. Information on available resources (services) is shared by periodically announcing availability of resources (services) through distributed tuple spaces. Allia [18] uses peer-to-peer caching and policy-driven agents to facilitate cross-platform service discovery.

4   Trust Management and Composability

4.1   Trust in Distributed Systems

Trust is an abstraction of individual beliefs and requirements that an entity has for specific situations and interactions. Creating a universally acceptable set of rules and mechanisms for specifying and reasoning about trust is a difficult process because of the variety in trust definitions. Researchers have defined trust concepts for many perspectives, with the result that trust definitions overlap or contradict each other [19]. There are numerous models of trust, although no rigorous classification of either trust or its models has been developed yet. Nevertheless, there is a subtle feature that differentiates a *generic trust model* [20,21,22,23] from a *trust management system* [24,25,26]; the former focuses on representing specific aspects of trust, such as authentication, reputation, and cooperation, whereas the second focuses on dynamically managing the lifespan of trust relationships.

Regardless of model or management system, there are a number of open problems dealing with the general concept of trust [19]. In this section we focus on a specific property of trust, essential in collaborative environments: serial composability of a chain of trust relationships. By definition, trust is not automatically transitive, yet effective models for composing trust are required in a number of cases. For instance, in some situations, the determination of trust is based on reputation while for other configurations trust evaluation must rely on cooperation bonds. The goal is to devise a systematic way for synthesizing the various trust models in a situation-aware framework. In addition to the general case, serial trust composition is required whenever a number of entities collaborate in executing a specific task.

4.2   Serial Trust Composability

In this case study, we restrict the scope of serial trust composability to an information sharing system that delivers information from a source to the intended recipients. Trust composability is illustrated in the next example. In any trust relationship, there is a *trustor* and a *trustee*. A trustor is the entity that places its trust in another entity to act as expected, within a particular *context*. This second entity is the trustee. A trust relationship is one-to-many when a group of trustees are trusted similarly within the same context. Current trust relationships are *pairwise* and support trust towards a non-interacting group of trustees. In Figure 2(a), A trusts B, C, and D to consume data *d* in a one-to-many relationship. Recognition of one group member as untrustworthy would not affect the trust placed in the remainder of the group, provided that the untrustworthy member is expelled.

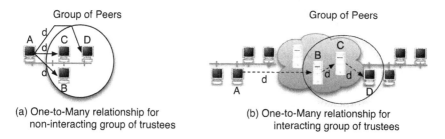

(a) One-to-Many relationship for   (b) One-to-Many relationship for
    non-interacting group of trustees   interacting group of trustees

Fig. 2. Pairwise and Composite Trust Views

A pairwise approach cannot encompass the complexity of trust in collaborations that go beyond two. Consider the WAN in Figure 2(b). A trusts D, that resides on a different LAN, to consume its data. Intermediate entities B and C forward this data to D, so some form of trust also exists between entity A and the forwarding servers. Malicious intermediary servers would affect the trustworthiness of data received by D. Untrustworthy servers cannot simply be expelled from the trust group but in many situations they must be replaced by trustworthy ones. Here, a trustor places its trust in interacting trustees that collaboratively execute a task rather than one alone. Such *composite* views of trust are implemented by systematic assessment of data trustworthiness when data is handled by a chain of different trustees.

In general, trust evaluation must integrate diverse inputs such as cooperation, competition, experience, recommendation, intrusion detection and assurance. Trust composition is a form of integration and it is aimed at reasoning about two risks: an information producer risks leakage or misuse of its information and an information consumer risks receiving inaccurate or malicious information for use. Composing trust in an information sharing system is founded on three concepts: trust expectations, information lifecycle decomposition, and information trustworthiness within its lifecycle. The methodical management of these concepts let us built individual pairwise trust relationships at different times during the information flow and examine a broader view of trust placed on the specific information stream.

Starting with the trust expectations concept, an entity forms expectations as a way to express concretely its interpretation of trust. Behavioral, security, and QoS requirements are included in a trustor's *expectations*. Expectations pertain to a specific trustor, in a given context, and this permits setting expectations based on individual perspective. An expectation includes all behavioral (competence and motivation), security, and QoS requirements derived from a trustor's goals, standards, principles, and morals.

The second concept deals with the information lifecycle. Information lifecycle is defined as the interval during which information is created and consumed. This interval is decomposed into three stages: generation, dissemination, and consumption. The entities responsible for the information at each stage are the information producer (generation stage), information dissemination medium (dissemination stage), and finally, information consumer (consumption stage). After decomposition, trust can be examined and evaluated at this finer granularity for each stage in the information lifecycle. *Pairwise trust* is the subjective and dynamic belief placed by an entity (trustor) on another entity (trustee) to act as expected during an information lifecycle stage.

There are two specialized forms of pairwise trust: *Information Provider Trust* (IPT) and *Information Consumer Trust* (ICT). IPT refers to the subjective and dynamic belief placed by an information consumer (trustor) on an information producer (trustee) to provide information as expected. Similarly, ICT refers to the subjective and dynamic belief placed by an information provider (trustor) on an information consumer (trustee) to consume information as expected.

Fig. 3. Pairwise Trust Relationships within Information Lifecycle

Finally, the third concept is that of information trustworthiness. Information trustworthiness during a particular lifecycle stage depends on the trustworthiness of the entity that handles the information during that stage. As a result, information trustworthiness is related to the trustworthiness of all entities that handle it, not just the creator or consumer of the data. Based on this principle, a composite view of trust is defined. Composite view of trust is the composition of subjective and dynamic beliefs placed by an actor (trustor) in other actors (trustees) to act as expected during the information lifecycle. In other words, composing trust is the synthesis of pairwise trust relationships.

Figure 3 illustrates the four basic pairwise trust relationships in a generic information sharing system, which is viewed from the perspective of its three required entities: a producer, an information dissemination medium, and a consumer. Synthesizing the pairs of pairwise relationships gives a composite view of trust, which is essential in managing the risk of information leakage and use of malicious information.

Composing trust becomes more complicated when the dissemination medium consists of a network of interconnected servers. Consider a publish-subscribe system, with publishers, subscribers, and event servers as the main entities. Among these entities there are three information flow lifecycles: publication advertisement from the publisher to the event servers, subscription request initiated by the subscriber and received by the event servers, and message forwarding from the publisher to the subscriber.

We focus on the message forwarding information flow lifecycle that has trust requirements as follows:

1. The publisher must be able to infer that the subscribers will not leak confidential information.
2. The publisher must be able to rely on the event servers regarding message forwarding.
3. The subscriber must be able to infer that publishers publish trustworthy data.
4. The subscriber must be able to rely on the event servers regarding message delivery.
5. Each event server must be able to rely on other trustworthy servers that it receives messages from or forwards messages to.

Figure 4 illustrates the pairwise trust relationships that satisfy the requirements mentioned above. It is important to note that each event server acts as both a producer and a consumer of information by forwarding to and receiving messages from its adjacent servers. Publishers and subscribers need the tools to synthesize the individual trust relationships for the servers involved in the forwarding operation so as to derive the trustworthiness of the information dissemination medium as a single entity. For example, ICT(Producer,Dissemination Medium) translates into the composition of the trustworthiness of the chain of the servers on the information delivery path.

4.3   Trust and the Electric Power Grid

An electric power grid consists not only of a network of generators, transmission lines, and distribution infrastructure to customer premises but is overlaid with a communications and control system which enables the economical and stable operation of the grid. The communications infrastructure for the electricity grids in Europe and the US are based on conceptual designs from the 1960s and have evolved very little since. In recent decades forces including industry restructuring, a lack of investment in transmission capacity, and almost no investment in research and development have converged to stress the grid and to highlight the inadequacy of its communications infrastructure [27]. This rudimentary infrastructure must match supply and demand and control

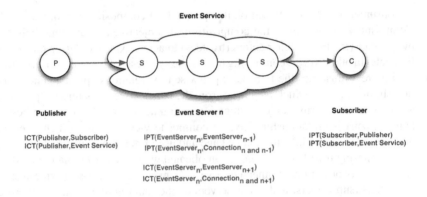

Fig. 4. Pairwise Trust Relationships in Publish-Subscribe Systems

grid-wide dynamics in real time with control mechanisms and sensors that are largely local in scope, and its limitations have been a factor in every recent major blackout. Its shortcomings also limit the deployment of better control and protection schemes [28].

We have been developing the GridStat framework (www.gridstat.net) to provide better communications services for the electric power grid. GridStat specializes the publish-subscribe paradigm for the semantics of status dissemination, and features QoS management. GridStat is involved in collaborations with a local electric utility and a national energy laboratory in the US. Through projects such as GridStat, the power grid's existing communication infrastructure is thus in the initial stages of being transformed into one with distributed services that must support composability in the ways described in Section 2. However, operating in a collaborative yet competitive environment gives rise to new challenges, especially in the realm of trust management, involving proper and legal disclosure of data between grid participants. Universally accessible information may result in compromising consumer privacy or extracting consumption production patterns, or even aiding in launching malicious cyber or physical attacks. Information passed by a competitor or the ability to delay a competitor's information may also provide significant and unfair competitive advantage. Uncontrolled data accessibility affects the stability of the grid from operational and financial perspectives.

Consider a simple early warning system that spans a number of regional districts. One of the important requirements for maintaining grid reliability and stability is to respond in a timely manner to various problems, including security attacks and operational failures. If a warning system were in place, cooperating power companies would have disclosed leading indicators of problems (virus information, status data involving local instabilities and perturbances not visible outside a given power company, etc.) to prevent further escalation of the failure with catastrophic consequences (e.g. bringing down the power grid, or part of it). One major practical difficulty in deploying such a system lies on the fact that power utilities are reluctant to share information that might jeopardize their business, either legally or financially. The source of their hesitation is their inability to quantify the risk of business discontinuity. Hence, in order to protect themselves financially and legally, utilities choose not to share information.

However, a utility's 'no share' policy can be relaxed in many cases if a trust management system provides automated predictions about the risk factors regarding interactions with other grid participants. For example, the disclosure of an indicator that is not market sensitive, under normal conditions, pertains no significant risk for the utility. Utilities do not trust each other, but, with adequate trust management support, sharing does not have to be a binary decision. Even though utilities need to restrict who has access to its published data, selective data sharing is still a possibility. The real challenge is to understand what trust means at a particular situation. Deciding what to share and with whom is a function of the trustworthiness of a utility at the other end of the interaction. Risk management is a vital element of decision making, as there is a cost for trusting but at the same time there is a cost for not trusting.

5   Discussion

Composability is a key property that has many facets. In this paper we provided a refined definition of composability that includes vertical composability across layers of mechanisms and services and horizontal composability between peers entities in these layers. We then examined composability issues and mechanisms in two different middleware frameworks for critical infrustructure services.

The examples presented in Section 3.2 for composable middleware services and application services illustrate the need for reconfigurable services to determine proposed composition and configurations. In the first two examples the different components of the Ad Hoc InfoWare middleware need to compose their functionalities in order to perform. The first one indicates how to create the complete DENS configuration in a resource poor network, and the second one shows how resource management improves when it can use high level information from KM. The third and fourth examples indicate how service and resource composition can help applications.

The examples involving trust management issues in the electric power grid presented in Section 4.3 involve composability in very different ways. The existence of multiple entities, with possibly diverse trust policies and mechanisms, makes trust composition for information flows that span these domains more complicated and less tractable. One of the lessons learned from decades of research in security is that the problem of how to compose two or more secure components into a secure system is hard and still remains an open problem. Trust composability requires synthesis of trust relationships. Unlike secure components, trust relationships' semantics are simpler and it is our opinion that it is feasible to compose chains of realistic relationships, given that these relationships are formally defined. While this is still open research, we believe that heuristic approximations and policy language support provide relaxations for many practical cases.

The composability techniques, issues, and tradeoffs presented in these two examples apply at least in part to other contexts. Most research projects and commercial products in the area of resource management are tailored for local area networks or at most a corporate enterprise scope, and thus do not have to deal with the severe availability issues that resource management for ad hoc mobile networks necessarily entails. However, as they become deployed in wide area networks—with their higher variations in both resource availability and the usage patterns, QoS requirements, and the sheer number of ubiquitous client applications—then the issues such resource management frameworks face become similar to issues faced by ad hoc mobile networks. The trust management issues and techniques described in this paper generalize beyond the electric power grid. Many environments where there are economic markets involved and real-time status being monitored will require such composability, including non-electric energy markets. However, we believe that many other kinds of distributed applications over the next decade will require serial composability of trust relationships, and corresponding trust infrastructures to provide them. Examples of such applications include so-called "virtual corporations" lashed together for a short or medium period of time to cooperate on a product or contract, or emergency response.

Acknowledgments

This research was funded in part by Norwegian Research Council in the IKT-2010 Program, Project Nr. 152929/431, and by Grant CCR-0326006 from the US National Science Foundation. We thank the members of the Ad Hoc InfoWare project team. We also thank Erek Göktürk and Ryan Johnston for their valuable feedback on this paper.

References

1. Zinky, J.A., Bakken, D.E., Schantz, R.E.: Architectural support for quality of service for CORBA objects. Theory and Practice of Object Systems **3** (1997) 55–73
2. Venkatasubramanian, N.: Safe composability of middleware services. Communications of the ACM **45** (2002) 49–52
3. Werner, M., Richling, J., Milanovic, N., Stantchev, V.: Composability concept for dependable embedded systems. In: Proceedings of the International Workshop on Dependable Embedded Systems at the 22nd Symposium on Reliable Distributed Systems (SRDS 2003), Florence, Italy (2003)
4. Milanovic, N., Malek, M.: Architectural support for automatic service composability. unpublished manuscript, available at http://www.informatik.hu-berlin.de/ milanovi/scc2005.pdf (2005)
5. Mishra, S., Schlichting, R.: Abstractions for constructing dependable distributed systems. Technical Report TR 92 -12, University of Arizona (1992)
6. Schantz, R., Schmidt, D.: Middleware for Distributed Systems - Evolving the Common Structure for Network-centric Applications. In: The Encyclopedia of Software Engineering. John Wiley and Sons (2001) 801 – 813
7. Chandra, P., Chu, Y.H., Fisher, A., Gao, J., Kosak, C., Ng, T.E., Steenkiste, P., Takahashi, E., Zhang, H.: Darwin: Customizable resource management for value-added network services. IEEE Network **15** (2001) 22–35
8. Skjelsvik, K.S., Goebel, V., Plagemann, T.: A highly available distributed event notification service for mobile ad-hoc networks. In: ACM/IFIP/USENIX 5th International Middleware Conference (Middleware 2004), Toronto, Canada (2004)
9. Perkins, C., Belding-Royer, E., Das, S.: RFC 3561: Ad hoc on-demand distance vector (aodv) routing (2003)
10. Phanse, K.S., DaSilva, L.A., Midkiff, S.F.: Design and demonstration of policy-based management in a multi-hop ad hoc network. Ad Hoc Networks, Elsevier Science (2003)

11. Cardei, I., Varadarajan, S., Pavan, A., Graba, L., Cardei, M., Min, M.: Resource management for ad-hoc wireless networks with cluster organization. Journal of Cluster Computing in the Internet, Kluwer Academic Publishers **7** (2004) 91–103

12. Lee, S.B., Ahn, G.S., Zhang, X., , Campbell, A.T.: INSIGNIA: An ip-based quality of service framework for mobile ad hoc networks. Journal of Parallel and Distributed Computing, Special issue on wireless and mobile computing and communications **60** (2000) 374–406

13. Ahn, K.M., Kim, S.: Optimal bandwidth allocation for bandwidth adaptation in wireless multimedia networks. Computers and Operations Research, Elsevier Science **30** (2003) 1917–1929 ad hoc, wireless networks hotspots, mutimedia.

14. Pei, G., Gerla, M.: Mobility management for hierarchical wireless networks. Mobile Networks and Applications archive, Kluwer Academic Publishers **6** (2001) 331–337

15. Chen, K., Shah, S.H., Nahrstedt, K.: Cross-layer design for data accessibility in mobile ad hoc networks. Special Issue on Multimedia Network Protocols and Enabling Radio Technologies, Kluwer Academic Publishers **21** (2002) 49–75

16. Li, B., Wang, K.H.: NonStop: Continuous multimedia streaming in wireless ad hoc networks with node mobility. IEEE Journal on Selected Areas in Communications **21** (2003) 1627–1641

17. Storey, M., Blair, G., Friday, A.: MARE: resource discovery and configuration in ad hoc networks. Mobile Networks and Applications **7** (2002) 377–387

18. Ratsimor, O., Chakraborty, D., Joshi, A., Finin, T.: Allia: Alliance-based service discovery for ad-hoc environments. In: Proceedings of the 2nd international Workshop on Mobile Commerce (WMC 2002), Atlanta, Georgia, USA (2002)

19. University of Southampton and QinetiQ: Trust Issues in Pervasive Environments. (2003)

20. Zimmermann, P.R.: The official PGP User's Guide. MIT Press (1995)

21. Marsh, S.: Formalizing Trust as a Computational Concept. Department of Computer Science, University of Sterling. (1994)

22. Josang, A.: Prospectives of modeling trust in information security. In: Proceedings of the 2nd Australasian Conference on Information Security and Privacy, Sydney, Australia (1997)

23. Abdul-Rahman, A., Hailes, S.: Supporting trust in virtual communities. In: Proceedings of the 33th Hawaii International Conference on System Sciences (HICSS), Maui, Hawaii (2000) 1769–1777

24. Blaze, M., Feigenbaum, J., Keromytis, A.D.: Keynote: Trust management for public key infrastructures. In: Proceedings of the 6th International Workshop on Security Protocols, Cambridge, UK (1998)

25. Sun Microsystems: Poblano: A Distributed Trust Model for Peer-to-Peer Networks. (2000)

26. Grandison, T., Sloman, M.: A survey of trust in internet applications. IEEE Communications Surveys and Tutorials **4** (2000)

27. Hauser, C.H., Bakken, D.E., Bose, A.: A failure to communicate: Next-generation communication requirements, technologies, and architecture for the electric power grid. IEEE Power and Energy **3** (2005) 47–55

28. Tomsovic, K., Bakken, D.E., Venkatasubramanian, M., Bose, A.: Designing the next generation of real-time control, communication and computations for large power systems. Proceedings of the IEEE (Special Issue on Energy Infrastructure Systems) **93** (2005) 965–979

Proof-Based System Engineering Using a Virtual System Model

Martin Biely[1], Gérard Le Lann[2], and Ulrich Schmid[1]

[1] Technische Universität Wien, Embedded Computing Systems Group E182/2
Treitlstraße 3, A-1040 Vienna, Austria
{biely,s}@ecs.tuwien.ac.at
[2] INRIA Rocquencourt, Project Novaltis
Domaine de Voluceau BP 105, F-78153 Le Chesnay Cedex, France
Gerard.Le_Lann@inria.fr

Abstract. This paper provides an overview of Proof-Based System Engineering (PBSE), which aims at improving the current practice of developing computer-based systems. PBSE is of particular relevance for safety critical applications and other systems where dependability properties are essential. This is particularly the case for applications in the aerospace domain targeted in the EC FP6 Integrated Project ASSERT. Applying PBSE both permits to eliminate most common design faults before embarking on the development of a system and maximizes reuse, which leads to significant savings in time and budgets. Particular emphasis is put on the requirements capture phase of PBSE, where a virtual system model is used as a novel means to structure the information to be captured.

1 Introduction

Stringent requirements for high availability, high reliability and safety in mission-or/and life-critical applications entail specific and complex constraints on the design, verification and validation of *computer-based systems* (CBS). The challenges thus involved are addressed by the *Proof-Based System Engineering* (PBSE) method, which builds upon INRIA's TRDF method ("Traitement Distribué", "Temps Réel", "Tolérance aux Fautes"), a generic method that has already been applied successfully in a number of former projects [1,2]. PBSE is currently applied in the FP6 Integrated Project ASSERT.[1]

Unlike most software engineering approaches, PBSE targets the entire CBS of an application, not just the software part of its constituents' embedded systems. Examples are the worldwide distributed CBS for a bank or — in the case of ASSERT — the CBS that spans spacecraft, the International Space Station, and

---

[1] ASSERT (IST-004033) is an IST-FP6 Integrated Project sponsored by the European Commission under the strategic objective of "Embedded Systems". Coordinated by the European Space Agency (ESA), the consortium consists of 29 partners from both academia and industry. Consult www.assert-online.org for further details.

M. Malek, E. Nette, and N. Suri (Eds.): ISAS 2005, LNCS 3694, pp. 164–179, 2005.

ground stations. Traditional formal/informal software engineering methods are primarily concerned with *how to build the specification right*, i.e., how to correctly implement some given specification. PBSE is orthogonal to these methods as it addresses the issues involved with *how to build the right specification*, which consists of:

- *building an adequate specification of the problem(s) to be solved*, by mandating a dedicated requirements capture phase prior to any system design, validation and implementation work,
- *building a correct specification of the solution(s)*, with a priori and maximum reusability of efforts, by mandating "forward" proofs in every step of the solution design, rather than "backwards" verification and testing.

PBSE focuses entirely on the CBS-centric non-functional requirements "hidden" in an application, however. It thus actually allows to separate functional requirements (application semantics) from non-functional requirements [3]: Application programmers, who may use standard formal/informal software engineering methods[2], can safely ignore non-functional aspects during functional analysis and design. PBSE experts, on the other hand, can abstract away functional requirements in the course of their work, which rests upon splitting the non-functional requirements into a set of *models* that specifies assumptions about the CBS's environment, and a set of *properties* that specifies desired system-level services and their QoS. The system-level solutions developed according to PBSE principles will guarantee that the CBS satisfies those properties in any environment that matches the assumptions stated in the models.

One of the primary purposes of PBSE is to eliminate faults made in the early phases of the overall life cycle: It is well known that faults made in the course of requirements capture phases are the dominant causes of project setbacks or operational failures, hence the major contributors to inflated costs and project overruns. Another primary purpose of PBSE is to reduce the complexity of the system integration and final testing phases, phases which are not well mastered under current practice. Finally, PBSE aims at composability checking, targeting the reuse and composition of designs and proofs, not just the reuse and composition of software or hardware components.

This paper provides an overview of the rationale and life cycle of PBSE, and introduces the virtual system model as our primary means to structure the requirements capture phase. It is organized as follows: The rationale for the need of a proof-based approach and some related work is given in Section 2. An overview of the PBSE life cycle, with particular emphasis on the PBSE requirements capture phase, is contained in Section 3. Section 4 provides a short example of the reuse possible with PBSE. The definition and usage of the virtual system model is presented in Section 5. A concluding discussion of PBSE in Section 6 completes the paper.

---

[2] Using formal SW engineering methods puts you in the desirable situation of having a continuous chain of proofs from problem specification to implemented solution.

2   Why PBSE ?

Consider critical systems, where criticality is related to the possible loss of life, mission or simply money. Obviously, such systems should be designed in a way that prevents such losses, or, more realistically, makes them sufficiently unlikely. For air traffic control systems, for example, it is required that system unavailability shall be less than 3 seconds a year, which translates into an availability figure of $1 - 10^{-7}$. With today's practice, however, too many of these systems fail, and too many projects are canceled, are late or more expensive than planned due to the difficulty of meeting such stringent requirements.

Software design & software development is commonly blamed for these problems, giving raise to the so-called "software crisis" in critical systems design. And indeed, as software became the dominating factor in today's computer-based systems, there is always some piece of software running when a system failure occurs. However, simply accusing software turns out to be wrong or at least misleading in many cases, since doing this ignores the difference between the cause of a system failure, i.e. the fault, and its observed manifestation, i.e. the failure.

In fact, several studies show that software is better than its reputation: For instance, an analysis of the causes of failures of the US public switched telephone network [4] shows *"that SW errors caused less downtime (2%) than any other source of failure except vandalism"*. Rather, overloads were recognized as the dominant cause (44%). Another example where software was blamed for a system failure is the well-known loss of the Ariane 5/501 launcher, which caused a financial loss in the order of 450 M€ and a 1 year delay for the Ariane 5 program. Although the inquiry board [5] concluded that poor software engineering practice was the culprit, other problems actually caused the failure [6,7].

Rather than in software [engineering], these and many other critical system failures have their roots in poor system engineering practice [8]. Of course this is not meant to suggest that the computer industry does not have problems in the field of software engineering, but rather that there are other areas (i.e., system engineering) that are even less mastered and have not received enough attention.

One major reason of failure is related to the specification generation process: With formal software engineering methods, under some restrictions, it can be verified that specifications are implemented correctly. But where do these specifications come from? It does not help much to be provided with a software component "proved correct" vis-à-vis its specification, if that specification is inappropriate ("incorrect") for the application/system problem considered. Proper requirements engineering methods [9,10,11,12] must be utilized to provide an agreed-upon specification of the problem to be solved. In general, however, this is difficult due to the inevitable intertwining of requirements and solutions [13] and the often conflicting requirements of different stakeholders [14]. In the context of ASSERT, the problem is further exacerbated by the difficult fault-tolerant distributed real-time computing problems typical of aerospace applications.

Another major problem is the level of complexity involved with proving systems-in-the-large [3,15]: Even a locally verified system component can suffer from inconsistencies and hidden non-functional dependencies with respect to other components in the system. So if such components, which behave correctly when run in isolation from each other, are executed together within a system, they could suffer from undesired interference and hence fail. The resulting failure is observed in the execution of the software, but the actual fault is rooted within poor system engineering practice: Global verification would have spotted system-level inconsistencies and hidden non-functional dependencies. Unfortunately, however, such techniques suffer from well-known state explosion problems and are hence infeasible for most real-world-size problems. Moreover, they are necessarily "a posteriori" verification approaches, which do not a allow the development of solutions that are correct-by-construction.

PBSE is the only method we are aware of that addresses these challenges in a common framework: PBSE/TRDF shares some of the goals of the Design-by-Contract approach [16] and the B method [17], notably the mandated use of non-ambiguous specifications and the fulfillment of proof obligations. However, PBSE addresses system-level concerns, regardless of the implementation technology resorted to in fine, rather than software-related concerns only. In the remainder of this paper, we will try to shed some light on how this is accomplished.

3   The PBSE Life Cycle

Before giving an overview of the phases of the PBSE life cycle, we need to introduce some basic notations. As mentioned above PBSE is concerned with building the correct specification of the problem to be solved, as well as building the correct specification of the solution. A *specification of a problem* will be denoted $\langle Z \rangle$, with $\langle z \rangle$ denoting the set of unvalued variables in $\langle Z \rangle$. The *Design specification of a system solution* will be denoted $[S]$, with the set of unvalued solution variables $[s]$ that correspond to the unvalued problem variables $\langle z \rangle$. Typical examples of such unvalued variables are process sets, deadlines, worst-case execution times, invariants for logical safety, density of failure occurrences. Note that the size and type of $\langle z \rangle$ and $[s]$ reflect the genericity of the specification of the problem $\langle Z \rangle$ and the solution $[S]$, respectively. The design specification $[S]$ is referred to as specification of a solution, because its implementation is the solution, denoted S, of the problem stated in $\langle Z \rangle$. In ASSERT we do not consider a specific mission, but rather (two) families of missions, resulting in two very generic pairs $\{\langle Z \rangle, [S]\}$ of problem and corresponding solution specification, which are referred to as *System families (SF)*.

A problem specification $\langle Z \rangle$ actually comprises two sub-specifications:

- *Models* $\langle m.Z \rangle$, which stipulate operational, technological, and environmental assumptions. They specify the *adversary* (Adv) for (the designers of) $[S]$.
- *Properties* $\langle p.Z \rangle$, which stipulate the desired services and QoS. They must be guaranteed by the operational system S (assuming $[S]$ is implemented correctly) in the presence of an adversary no stronger than $\langle m.Z \rangle$.

Specifications such as $\langle Z \rangle$ are written in restricted natural language: All terms in $\langle Z \rangle$ must have formal or technical definitions in scientific or engineering dis-

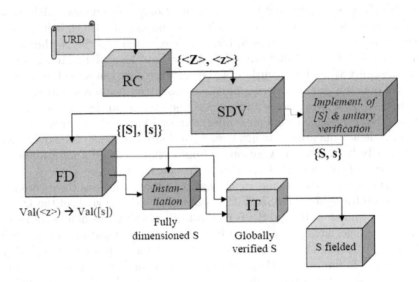

Fig. 1. *Schematic representation of the entire PBSE life cycle*

ciplines (computer science terminology, mathematics, etc.), to the exception of conjunctions, articles, and other syntactic elements. Examples include:

- "distributed" ≡ "current global state cannot be known",
- "serializable execution" ≡ "interleaved execution identical to some sequential execution",
- "Byzantine" ≡ "arbitrary behavior".

Figure 1 shows a schematic representation of the entire PBSE life cycle. Phases that are proper to PBSE are the RC, SDV, FD and IT phases. The RC and SDV phases precede the instantiation of [S], the FD phase precedes the instantiation of every specific customized release of S, and the IT phase serves to derive automatically the suite of tests needed to conduct the integration testing (global verification) of S. The implementation of [S] and unitary verification, on the other hand, are fully within the realm of formal/informal software engineering.

Therefore, the PBSE process spans all RC, SDV, FD and IT life cycle phases whenever a novel problem Z is considered and some solution S is to be fielded. Conversely, after a pair $\{\langle Z\rangle, [S]\}$ has been constructed, only the FD and the IT phases need be conducted for the fielding of some specific release of S. Customized releases are obtained by assigning values to free problem variables in $\langle z\rangle$ and running the FD phase, which produces values for the free system variables in $[s]$.

3.1   The Requirements Capture Phase

The *Requirement Capture* (RC) phase bridges the gap between the application-centric requirements and the resulting CBS-centric requirements. The input of the RC phase is a document, called *user requirements document* (URD) in the sequel, which describes the objectives of the application in the client's domain-specific terminology. The result of the RC phase is a specification of the system-level computer-based problems/requirements $\langle Z \rangle$, which matches the URD.

In the RC phase, the application/mission requirements are separated from reuse considerations — PBSE accommodates the mandatory use of pre-existing partial solutions (e.g., COTS products). Design concerns, related to how $\langle Z \rangle$ could be solved and/or some [S] implemented, are totally ignored.

An existing or novel component of a to-be-designed system is called an *entity*. The modeling of entities is done similar to the I/O automata formalism [18]:

- Inputs are (specifications of) incoming events and associated shared data, arrival laws (loads), failures,
- internals are (specifications of) processes (structure, worst-case execution times) and shared data/states,
- outputs are (specifications of) outgoing events and associated shared data, failures.

These models are intrinsic to a given entity. Inputs and outputs correspond to behaviors in I/O automata. Properties serve to specify desired properties, which may differ from (intrinsic) outputs.

Operationally, requirements capture is a two step process. In step 1, models and properties are captured on a per entity/level basis, in strict isolation of each other. This work can be done by multiple teams in parallel, for different entities or collections of entities. Since the collective behavior of sets of entities (e.g. multiple programs multiplexed over a CBS) is usually also relevant, the desired properties for such sets may also be captured. An example would be the "serializability" property [19] for a set of application programs that share updatable and persistent data. Finally, since computer-based systems are never built from scratch in real projects, it is possible to specify, at RC time, which pre-existing components (hardware, software) are to be reused. To be part of the proof chain that spans from $\langle Z \rangle$ to [S], however, a reused component E must have a companion technical leaflet (see Section 3.2) that also includes its $\langle Z(E) \rangle$.

In the second step of the RC phase, every entity E is revisited, considering all models captured at the end of step 1 which are appropriate. For example, some failure models and failure occurrence models have been captured (during step 1) for processing entities (abstractions of processors). Causes of such failures are cosmic rays, vibrations, and so on. Separately, some failure models and failure occurrence models have been captured (during step 1) for application entity E (abstraction of a software/functional process). Causes of such failures are software design and implementation faults. During step 2, entity E is "revisited", in order to specify its intrinsic behaviors in the presence of failing processing entities (ignored at step 1).

In the absence of tools[3], the RC phase is typically performed via interactive meetings, where the stakeholders scan the application's URD according to the PBSE *RC Guide*. The RC Guide is a menu of classes of models and of properties orthogonal to each other, that constitute a multidimensional space Π. The 10 classes that define Π (6 model classes, 4 property classes) are as follows:

- Computational Models, Resource and Data Models, Process Models, Event and Event Arrival Models, Failure Models, Failure Occurrence Models
- Logical Safety, Liveness, Timeliness, Dependability properties

Any specification $\langle Z \rangle$ corresponds to a region within Π.

It is this process that makes it possible to capture the properties $\langle p.Z \rangle$ and the adversary $\langle m.Z \rangle$ for the entire CBS. Thanks to $\langle Z \rangle$, it is then possible to detect some impossibility results at the RC stage (in addition to incompleteness, over-specification, etc.). Since those problems are found very early in the life cycle, in particular, before any design, implementation and testing work has been done, this distinguished PBSE feature considerably saves time and money. The SDV phase is in fact entered only when some $\langle Z \rangle$ that is free from obvious impossibility results has been established.

3.2   The System Design and Validation Phase

The other PBSE phase that occurs before any implementation work on the system is the *System Design and Validation* (SDV) phase, which aims at building the specification [S] of a solution S that provably solves the problem(s) captured in $\langle Z \rangle$. The outcome of the SDV phase is a *technical leaflet* (TL) for pair $\{\langle Z \rangle, [S]\}$, which is a 5-tuple $\{\langle Z|z \rangle, [S|s], \text{proofs}, \text{Cs}, \text{FD\_Oracle}\}$ consisting of

- the problem specification $\langle Z \rangle$, with unvalued variables $\langle z \rangle$,
- the solution specification [S], with unvalued variables [s] (which match $\langle z \rangle$), usually resting upon some *design assumptions* (DA),
- *proofs* (or pointers to such proofs) that [S] meets $\langle Z \rangle$,
- *feasibility conditions* (FCs), i.e., analytical conditions that must hold between $\langle z \rangle$ and [s] in order to ensure that S's valued properties (e.g. response times and availability figures) hold,
- *FD_Oracle*, (the specification of) a computer program that instantiates the FCs in order to simplify and speed-up the FD phase; the FD_Oracle can be developed any time after completion of the SDV phase.

PBSE does not make any requirements on how the properties and models are expressed. Typically, however, Logical Safety properties are expressed as invariants defined over values taken by sets of variables which represent the state of the spacecraft (or more generally, of the CBS). Proofs for Logical Safety or

---

[3] For the past decade, PBSE/TRDF has been applied without tool support. One of the goals of ASSERT is to develop the prototype of a RC tool (SDV and FD tool prototypes as well), whereby the RC work conducted manually at the beginning of ASSERT would be replayed in a somewhat automated manner.

Liveness are proofs in logic. Timeliness proofs are proofs in Combinatorial Analysis/Scheduling Theory. Dependability proofs are combinations of such proofs, augmented with coverage analysis. In $\langle p.Z \rangle$, one finds $Cov([S])$, which stands for the smallest acceptable coverage to be met by [S], as stipulated by the client.

The Technical leaflet for the whole CBS system or the system family — that is for the pair $\{\langle Z \rangle, [S]\}$ — is inevitably a set of TLs, one for each of its components, which can be either *application-centric building blocks* (ABBs) or *computer-centric building blocks* (CBBs). Typical CBBs deal with system-level issues like distributed resource management, failure detection and/or masking, synchronization, concurrency control, timeliness, etc, whereas ABBs realize the actual functional requirements. CBBs provide the abstraction of a computing environment as "perfect" as required for the ABBs. Quite often, "perfection" means keeping invisible such things as concurrent computations or failures. Application programmers, who design ABBs, can hence concentrate solely on application-related issues, using their favorite formal/informal software engineering methods, and need not worry about specific peculiarities or/and imperfections of the underlying CBS. Moreover, ABBs can be developed and verified in isolation of each other, with no (or quite limited) need for global or integrated verification.

The specification of the solution [S] is in fact a modular specification, which results from building a *design tree* rooted at $\langle Z \rangle$: Top-level ABBs are typically just "containers" for application-level functionality. The DAs of such top-level ABBs are hence fairly idealistic, like "there are only perfect processors in the system". Since such assumptions have a rather bad coverage, this leads to corresponding (sub-)problem specification(s) to be met by the specifications of novel or reused CBBs, which must be dealt with at the next (lower) level of the design tree. This process of successive refinement proceeds along some number of branches. On a given branch, one stops designing whenever both (1) the specification arrived at is deemed implementable and (2) its DAs have a coverage at least as high as $Cov([S])$.

Another important output of the SDV phase are the FCs, which are typically a set of constraints required for the solution [S] to work. In order to simplify checking of the feasibility of some particular dimensioning of the problem variables $\langle z \rangle$ and calculating the corresponding dimensioning of [s], FCs are instantiated as a computer program (referred to as an *FD_Oracle*, valid for pair $\{\langle Z \rangle, [S]\}$), which can be developed any time after completion of the SDV phase.

3.3   The Feasibility and Dimensioning Phase

For any given problem $\langle Z \rangle$, the SDV phase leading to [S] — as well as *implementation & unitary verification*, which is not a PBSE activity — is conducted only once, i.e., [S] for $\langle Z \rangle$ needs to be established and proved only once.

By contrast, the *Feasibility and Dimensioning* (FD) phase (as well as the following *instantiation* phase and the IT phase) has to be conducted every time a specifically customized release of [S] is to be fielded. The FD phase consists in a user choosing some specific valuation $Val(\langle z \rangle)$ of the unvalued problem variables in $\langle Z \rangle$, running the FD_Oracle, and (if possible) obtaining the resulting valuation $Val([s])$ of the unvalued system/solution variables in [S]. For example, this is how one knows the smallest period of activation of a scheduler, the smallest memory space for a waiting queue, or the smallest degree of redundancy which are necessary for meeting the required reliability figures. If the FCs are violated, the FD_Oracle indicates the reasons why, in which case some of the values assigned to ⟨z⟩ must be "relaxed" (e.g., some deadlines augmented).

3.4   The Integration Testing Phase

As stated above, *implementation & unitary verification* is not a PBSE activity. However, in order to maintain a continuous chain of proofs, not only from ⟨Z⟩ to [S], but also from [S] to S, automatic code generation and formal (local) verification should be used also during implementation of S.

When implementation and unitary verification has been conducted for all BBs that are part of S, the *Integration Testing* (IT) phase can be performed in order to check whether the composition of BBs is correct — a daunting task under current practice, since exponential complexity is to be faced. This is not the case under PBSE, which eliminates the classic state explosion problem involved in global verification and improves the achieved coverage compared to current testing practices, respectively, for two reasons essentially:

- No or just some limited global verification is necessary (proofs replacing possibly huge sets of tests).
- The suite of tests to be performed can be generated (if so desired) as a by-product of running the FD_Oracle, rather than by "guessing" them.

Consequently, with PBSE, integration testing work is typically unnecessary or at least limited, since "composition correctness" has been proved during the SDV phase (otherwise, [S] would not exist).

4   A Design and Reuse Example

Among the major advantages of PBSE, which is also a major target of ASSERT, is its potential for re-using BBs specified and designed in former projects. Given that (1) PBSE is concerned about system-level problems, which appear over and over again in many different applications, and (2) reuse in PBSE also includes reusing the design specifications and the proofs, the potential for reuse is indeed high.[4] Thanks to the TLs, the common practice of developing everything from scratch and/or best-effort reuse of existing components can be replaced by a systematic reuse exercise according to PBSE principles, i.e. conditioned upon using provably correct compositions of components.

A major goal of ASSERT in this realm is the definition of *system families* (SF), which represent a reasonably large class of space applications that share a

---

[4] Consequently, the budget and time savings that can be achieved with PBSE are high as well.

sufficiently large set of common properties. Ideally, when a new application is to be developed from a system family, most of the family's generic BBs are reused, and only the few ones that encapsulate application-specific functionality need to be modified/added.

One of the pilot projects in ASSERT is devoted to the definition of a system family for satellite missions. It is based upon a generic specification $\langle Z(SF) \rangle$ that captures the CBS problem common to such missions. ASSERT shall end up with the design specification (+ implementation) of a solution [S(SF)], consisting of a set of generic BBs, that matches $\langle Z(SF) \rangle$. In order to develop a particular satellite mission, say, a telecommunications satellite (TS), which is known to belong to SF (since $\langle Z(TS) \rangle \equiv \langle Z(SF) \rangle$), a user simply decides on some valuation $Val(\langle z \rangle)$ mirroring the TS-centric instantiation of SF and runs the FD_Oracle that was built for $\langle Z(SF) \rangle$, [S(SF)]. In other words the user has to conduct the FD and IT phases only. If $\langle Z(TS) \rangle \not\equiv \langle Z(SF) \rangle$, then some SDV work is necessary, reusing the BBs developed for SF. Consequently, the design tree for TS quickly reaches nodes that are already available. As a consequence, real design and implementation work is only needed for features that are specific for TS.

More generally, assume that, at some node of the SDV tree rooted at $\langle Z(TS) \rangle$, one is contemplating the specification of a sub-problem $\{ \langle m.Z(X) \rangle, \langle p.Z(X) \rangle \}$, and that there is an existing TL matching this specification (searches for matching TLs will be done by an SDV tool in the future). Since the specification [S] found in this TL has been proved correct for $\langle Z(X) \rangle$, the corresponding solution can simply be reused as such (with or without prior dimensioning), provided the conditions for stopping the SDV work for that SDV tree node are met.

For example, consider that $\langle Z(X) \rangle$ is the specification of some problem that was addressed in the A3M project[5]. The A3M objective was to develop a new generation of generic components as basic building blocks for the development of middleware targeting various on-board space applications [20]. The core CBBs developed in A3M employ asynchronous distributed fault-tolerant algorithms [21] for distributed consensus, coordination, and atomic commit, which are built atop of Chandra/Toueg unreliable failure detectors [22]. They rest upon design assumptions such as processor crashes, arbitrarily variable delays, and reliable communications, but do not need any notion of global time in the system. Thus the logical safety and liveness properties stated in $\{ \langle m.Z(X) \rangle, \langle p.Z(X) \rangle \}$ hold with these CBBs regardless of the (implementation-dependent) timing properties of the underlying system.

When the design assumptions meet the conditions for stopping the SDV work, then A3M solution can be reused in ASSERT as such. If some design assumption, say, Y, does not meet the conditions for stopping the SDV work (e.g., if processor omission failures and/or unreliable communications are assumed for the ASSERT SF), then Y translates into a sub-problem $\{ \langle m.Z(Y) \rangle, \langle p.Z(Y) \rangle \}$ (e.g., simulating processor crashes in the presence of omissions and providing reliable communications over unreliable channels), and the SDV work continues.

---

[5] Advanced Avionics Architecture and Modules, conducted by EADS Astrium, INRIA, LAAS, Axlog Ingenierie and funded by ESA/ESTEC (2001–2003).

5   The Virtual System Model

The PBSE requirements capture phase poses some particular challenges in that it targets inherent application needs only, rather than (premature) design considerations. In fact, freezing requirements actually rooted in traditional or even anticipated solutions in $\langle Z \rangle$ unnecessarily restricts the solution space for the later SDV work. This problem became particularly apparent during the RC phase for the complex system families/pilot projects in ASSERT, which was conducted by multidisciplinary teams: Following industrial practice, and quite natural for engineering disciplines, the initial versions of the URDs were heavily populated with a priori chosen system architectures, failure management strategies, process synchrony assumptions and other design considerations. Extracting out exactly those requirements that must be fulfilled by a CBS in order to meet the demands of the particular application (but nothing else) turned out to be a challenging task, cf. [13].

5.1   Using the VS Model for RC

In order to alleviate this problem, we introduced the *virtual system model* (VS model) for requirements capture. The virtual system model consists of several levels, which represent different levels of abstraction of a computer-based system. A level is populated by entities that represent components of a CBS at the corresponding level of abstraction, i.e., can be seen as a suitable "projection" of a CBS onto some specific abstraction level. Consequently, the VS model can be employed for reasoning about a yet-to-be-designed system as well.

The levels foreseen in the VS model may be domain-dependent. In ASSERT, the following levels have been identified to be necessary and sufficient for embedded systems in the aerospace domain: *Equipment & Humans level* (EH-Level), *Application level* (AP-Level), *Middleware level* (MW-Level), *Basic Service level* (BS-Level), and *Hardware level* (HW-Level).

The EH-Level provides the highest level of abstraction. It "connects" a CBS with its environment. It is populated with entities that may or may not be considered part of the CBS. They include external equipment, sensors and actuators as well as human users. Although in the implemented system EH-Level entities are connected by means of the HW-Level, which encompasses the raw computing and communication hardware of a CBS, in the VS model the EH entities can directly interact with entities at any level.

The AP-Level is made up of all the entities that instantiate the application's semantics. They are distributed/partitioned according to functional analysis considerations.

MW-Level entities typically serve two purposes: First, they provide a level of encapsulation, which allows application-level entities to access resources in a way that is independent of their physical location. Second, they typically host all the distributed fault-tolerant algorithms and protocols needed to "solve" $\langle Z \rangle$, i.e., the system-level algorithms and protocols specified within [S].

The BS-Level is populated with entities that augment/encapsulate the raw services provided by HW-Level entities in a generic manner, in order to provide universal and elementary services needed by entities residing at higher levels. Typically, BS entities are operating systems, real-time kernels, link communication protocols, TCP-like communication protocols, I/O handlers, memory access/management protocols, etc.

Finally, the HW-Level is populated with entities which provide the physical capability to execute programs (SW, firmware, gate-level compiled code, etc.), and to exchange bits over physical communication entities (both on-board and long-haul communications, communications with sensors and actuators, etc.). In other words the HW-Level provides the raw "execution machinery", but does not include programs written in HW, such as the logic in gate level compiled code of an ASIC.

In fact, although the VS levels above appear to follow traditional implementation levels, it is important to understand that they are not meant to imply any particular implementation, since an AP-Level entity may actually be implemented as a real AP-Level SW process, or as a triple {AP-Level SW component, BS-Level SW component, HW-Level HW component}, and might even involve on-line reconfiguration. Moreover, VS levels do not have any particular hierarchical relationship. They must rather be viewed as sets of orthogonal entities that may have all kinds of mutual interactions: AP-Level entities in the VS model are not restricted to interact solely with the MW-Level in order to invoke services, nor do MW-Level entities provide services to the AP-Level only. For example, a BS-Level entity — and even a HW-Level entity — may invoke a service at the MW-Level.

Consequently, the VS model used for conducting a PBSE RC phase is generic, in the sense that it does not carry any restrictions relative to the construction of $\langle Z \rangle$ or future SDV work leading to [S]. Hence, it can indeed be employed for capturing $\langle Z \rangle$ for a yet-to-be-designed system in the PBSE RC phase.

The VS model opens up another level of "separation of concerns", beyond PBSE's ability to deal with an application's functional and non-functional aspects independently of each other: It allows to capture models and properties at every VS level independently and in strict isolation of each other. This leads to a significant reduction of the overall complexity of the PBSE RC phase and allows even further parallelization of the RC work, which reflects reality as experts in AP-Level software are most likely not experts in space-compliant hardware.

More specifically, the whole set of application requirements can be mapped onto (or rather: "sliced" according to) the different levels of abstraction corresponding to the VS levels. For every level, the resulting projections ("slices") can then be captured independently of the other levels. Note that properties are always associated with the level where they are required to hold. If, for instance, some AP level processes shall enjoy the ACID properties of transactions, the atomicity, concurrency, isolation, and durability properties [19] are captured for the AP-Level in $\langle p.Z \rangle$. Although those properties are likely to be provided by MW-Level concurrency control algorithms in the yet-to-be-developed solution

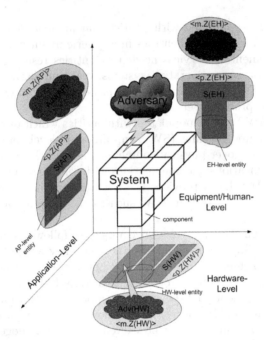

Fig. 2. *Visualization of the relation between system requirements and designed system projected onto VS levels: The designed system must provide properties that are fully within the specified properties when exposed to an adversary that is not stronger than the specified models*

[S], given the currently affordable technology, this shall not be frozen at RC time. Hence, the ACID properties, which are captured at the AP-Level, are not captured *for* the MW-Level. Any entity E of any level that has a TL showing that E provides the ACID properties is a correct BB for providing those properties at the AP-Level.

5.2   Using the VS Model for SDV

Figure 2 visualizes the resulting orthogonalization of the RC capture phase enabled by the VS model. Consider the AP-Level, for example. Let $\langle m.Z(AP)\rangle$ and $\langle p.Z(AP)\rangle$ denote the models and properties captured for this level, i.e., the projections of the whole set of requirements "hidden" in the application's URD onto the AP-Level. PBSE SDV work must eventually ensure that, whenever the HW-Level faces an adversary Adv(HW) that is not stronger than specified in $\langle m.Z(HW)\rangle$, the behavior of the forthcoming solution [S] (and hence the fielded system S) projected onto the AP-Level must stay within the behaviors stated in $\langle p.Z(AP)\rangle$.

Of course, the SDV work that provides the solution specification [S] must eventually consider all combinations of models captured in $\langle m.Z\rangle$ in order to consider the worst-case failure occurrences at all levels *simultaneously*, for example.

That is, in sharp contrast to the RC phase, where everything can be considered in isolation, the SDV phase must deal with the combinatorial complexity of combining the independently captured models and properties, including combined failure occurrences, worst-case event arrivals, etc.

From the virtual system model point of view in ASSERT, the most important levels for RC capture are the AP- and EH-Level, since entities at these levels are in fact the "end users" of the CBS. Note that, for the EH-Level, only the specification (models and properties) with respect to the interface(s) with the CBS is of concern.

If there was no reuse of pre-existing products, nothing would have to be captured at the levels below the AP-Level. Since systems are almost never implemented from scratch, however, the RC phase is also concerned with the identification of models and properties that characterize pre-existing BBs at any level, especially at the HW-Level and BS-Level. Any pre-existing product is either *trusted* or *not trusted*. By definition, at the time of writing, all existing products have been developed in some former projects *without* applying PBSE. Trusted products are those which have been developed and tested following certain rules, typically those stipulated by agencies or enforced by certification bodies (e.g., DO-178 or IEC standards, SILs in the UK).

Of course, although such products are considered trusted on the basis of careful design, diversified redundancy, sufficient testing, space-compliance and other means, this does not imply that they are fault-free. Nevertheless, the nature of the technical data available for a trusted product makes it possible to do some reverse PBSE work, i.e., to construct its TL a posteriori, at least partially (its models and properties, as well as its design assumptions). This way, pre-existing trusted products can be incorporated in $\langle Z \rangle$ and [S].

6   Concluding Discussion of PBSE

To achieve its ambitious goal of facilitating provably correct and reusable engineering work for critical fault-tolerant distributed real-time embedded systems, PBSE combines a number of different features in a common framework. First of all, a dedicated requirements capture phase has to be conducted, which provides an agreed-upon specification of the problem $\langle Z \rangle$ to be solved. $\langle Z \rangle$ not only describes what is to be achieved (properties $\langle p.Z \rangle$), but also under which conditions/circumstances (models $\langle m.Z \rangle$). The VS model has been introduced as an effective means to capture those properties and models in isolation of each other.

In general, conducting a requirements capture phase is difficult, for several reasons, such as: The intertwining of requirements and solutions, conflicting requirements of different stakeholders or the need for early freezing of requirements (waterfall model). PBSE does not suffer from those problems, however, for two reasons essentially: First, PBSE focuses solely on non-functional CBS-centric issues, which are reasonably independent of the particular application requirements. Second, the unvalued problem and solution variables $\langle z \rangle$ and [s] allow to introduce a user-decided degree of genericity in $\langle Z \rangle$ and [S], respectively, which effectively eliminates the well known problem having to nullify SDV work whenever $\langle Z \rangle$ is changed.

Finally, rather than on "a posteriori" global verification, PBSE rests upon "a priori" proof obligations to be fulfilled in the course of system design activities. PBSE hence necessarily saves time and money due to the fact that it provides solutions that are correct-by-construction; no time and money are wasted on trial and error-detection-and-correction iterations. Additionally, the concept of reuse also goes beyond what is commonly associated with this term: Not only existing implementations of BBs can be reused, but also their designs and proofs. Component reuse in PBSE is in fact similar in spirit to the use of existing lemmas for proving a new theorem in mathematics.

There is the widespread belief that proofs are too difficult to do in daily practice, a problem that has also slowed down the acceptance of formal software engineering methods. However, system engineers are not supposed to "do the proofs" (unless they run into a non-generic or unknown system problem). A fully developed PBSE process will eventually allow engineers and technicians to simply use instruction manuals, supported by appropriate tools, as is the case in other fields with good and mature engineering practice [23]: In handbooks for electricians, for example, one finds rules for how to install derivations, for computing voltages, etc. It is never the case that an electrician is asked to demonstrate anew the correctness of his doings based on Ohm laws or Kirchoff laws. Nevertheless, rules of good practice in mature engineering domains rest entirely upon such scientific results. We anticipate that this is going to happen to system engineering for computer-based systems as well: Rather than being founded on "experience, "intuition", or "good sense", rules of good system engineering practice will eventually rest upon science, and PBSE has been developed for reaching this goal.

References

1. Le Lann, G.: Proof-based system engineering and embedded systems. In: European School on Embedded Systems (Veldhoven, NL, Nov. 1996). Volume 1494 of Lecture Notes in Computer Science., Springer-Verlag Pub. (1998) 208–248 invited paper.
2. Le Lann, G.: Models, proofs and the engineering of computer-based systems: A reality check. In: Proc. 9th Annual Intl. INCOSE Symposium on Systems Engineering: Sharing the Future. Volume 4. (1999) 495–502 Best Paper Award.
3. Chung, L., Nixon, B.A., Yu, E., Mylopoulos, J.: Non-Functional Requirements in Software Engineering. Kluwer Academic Publishing (2000)
4. Kuhn, D.: Sources of the failure in the public switched telephone network. IEEE Computer **30** (1997) 31–36
5. Inquiry Board Report: ARIANE 5 — Flight 501 Failure. (1996) Available online http://ravel.esrin.esa.it/docs/esa-x-1819eng.pdf.
6. Le Lann, G.: An analysis of the ariane 5 flight 501 failure—a system engineering perspective. In: Proceedings of the IEEE International Conference and Workshop on Engineering of Computer-Based Systems. (1997) 339–346
7. Le Lann, G.: The failure of satellite launcher ariane 4.5. Safety Critical Mailing List at http://www.cs.york.ac.uk/hise/text/sclist/lelannariane.html Archived Contributions, Contribution on the Failure of Ariane 5 flight 501 (1999)

8. Zave, P., Jackson, M.: Four dark corners of requirements engineering. ACM Trans. Softw. Eng. Methodol. **6** (1997) 1–30
9. Jackson, M.: Software requirements & specifications: A lexicon of practice, principles and prejudices. ACM Press/Addison-Wesley Publishing Co. (1995)
10. Zave, P.: Classification of research efforts in requirements engineering. ACM Comput. Surv. **29** (1997) 315–321
11. Nuseibeh, B., Easterbrook, S.: Requirements engineering: A roadmap. In: ICSE '00: Proceedings of the Conference on The Future of Software Engineering, ACM Press (2000) 35–46
12. Jackson, M.: The meaning of requirements. Ann. Software Eng. **3** (1997) 5–21
13. Swartout, W., Balzer, R.: On the inevitable intertwining of specification and implementation. Commun. ACM **25** (1982) 438–440
14. Sabetzadeh, M., Easterbrook, S.: Analysis of inconsistency in graph-based viewpoints: A category-theoretic approach. In: Proceedings of the 18th IEEE International Conference on Automated Software Engineering. (2003) 12–21
15. Stevens, R., Brook, P., Jackson, K., Arnold, S.: Systems Engineering: Coping with complexity. Prentice Hall (1998)
16. Meyer, B.: Applying design by contract. IEEE Computer **25** (1992) 40–51
17. Abrial, J.: The B Book. Cambridge University Press (1996)
18. Lynch, N.: Distributed Algorithms. Morgan Kaufman (1996)
19. Bernstein, P., Hadzilacos, V., Goodman, N.: Concurrency Control and Recovery in Database Systems. Addison-Wesley Publishing Company (1987)
20. Honvault, C., Le Roy, M., Gula, P., Fabre, J.C., Le Lann, G., Bornschlegl, E.: Novel generic middleware building blocks for dependable modular avionics systems. In: Proceedings of the 5th European Dependable Computing Conference (EDCC-5). Volume 3463 of LNCS., Budapest, Hungary, Springer Verlag (2005) 140–153
21. Hermant, J.F., Le Lann, G.: Fast asynchronous uniform consensus in real-time distributed systems. IEEE Transactions on Computers **51** (2002) 931–944
22. Chandra, T.D., Toueg, S.: Unreliable failure detectors for reliable distributed systems. Journal of the ACM **43** (1996) 225–267
23. Maibaum, T.: Mathematical foundations of software engineering: A roadmap. In: ICSE '00: Proceedings of the Conference on The Future of Software Engineering, ACM Press (2000) 161–172

Evaluation of the Impact of Congestion on Service Availability in GPRS Infrastructures

Paolo Lollini[1], Andrea Bondavalli[1], and Felicita Di Giandomenico[2]

[1] University of Florence, Dip. Sistemi e Informatica,
viale Morgagni 65, I-50134, Italy
{lollini, a.bondavalli}@dsi.unifi.it
[2] Italian National Research Council, ISTI Dept.,
via Moruzzi 1, I-56124, Italy
digiandomenico@isti.cnr.it

Abstract. This paper deals with the congestion analysis of a GPRS infrastructure composed by a number of adjacent cells partially overlapped. We consider one cell as affected by an outage and through a transient analysis we evaluate the effectiveness of a specific class of resource management techniques for congestion treatment in terms of service availability related indicators. The classical availability analysis is thus enhanced, by taking into account the congestion following outages and its impact on user's perceived QoS, both in each cell and in the overall GPRS network. In order to efficiently solve the large and complex model capturing the network's behavior, we introduce a solution technique in which the solution of the entire model is constructed on the basis of the solutions of the individual sub-models.

1 Introduction

Congestion events constitute a critical problem in the operational life of networked systems. A network is congested when the available resources are not sufficient to satisfy the experienced workload traffic, and this can occur for many reasons, such as in case of extraordinary events determining an increase of traffic, or in case of unavailability of some network resources because of malfunctions (outage). Careful management techniques are necessary, to alleviate the consequences of such phenomena. The IST-2001-38229 CAUTION++ project [1] aims at building a resource management system to efficiently cope with congestion events in heterogeneous wireless networks. Management techniques are usually equipped with internal parameters, whose values have to be properly assigned in accordance with the specific system characteristics. In order to support this "fine-tuning" activity, a model-based analysis is promoted in CAUTION++ to analyze the behavior of the management techniques and to understand the impact of techniques and networks configuration parameters on properly identified Quality of Service indicators.

In this work, the focus is on the General Packet Radio Service (GPRS) technology, which has been already analyzed in previous studies under more simplistic network configurations. An inspiring work is certainly [2], in which the

M. Malek, E. Nette, and N. Suri (Eds.): ISAS 2005, LNCS 3694, pp. 180–195, 2005.

authors analyze the dependability of a GPRS cell under outage conditions. Another work ([3]) evaluates the effects of outage periods on the service provision considering two GPRS cells partially overlapping (and then possibly interacting), and accounting for outage congestion treatment and outage recovery.

In this paper, we perform a major extension and refinement to the previous studies, by setting up a modeling framework able to deal with a general GPRS infrastructure, where clusters of cells are considered, each cluster being realized through a number of partially overlapping cells. In case of an outage experienced by a cell in a cluster, a Resource Management Technique (RMT) is put in place to alleviate the congestion in the affected cell by distributing part of its traffic (users requests) on all the neighbor cells. In such a system context, we propose a methodology to evaluate the impact of congestion treatment on all the cells. The purpose of such analysis is to provide feedbacks for an optimal tuning of the parameters of the RMT (namely, the number of users to switch), so as to have the highest efficacy from its application towards resolving the congestion event. The definition of the general framework for the analysis of GPRS infrastructures has required a relevant effort, especially in the evaluation phase, due to the high level of complexity that can lead to very large state spaces for state-based analytical solutions or unacceptably long solution times for simulations. In order to efficiently solve the large and complex model capturing the network's behavior, we introduce a solution technique that follows a "divide and conquer" approach, in which the solution of the entire model is constructed on the basis of the solutions of the individual submodels.

In the literature, many works tried to master complexity developing new techniques to solve models. [4] details some techniques for generating and solving large state-space representations of models. In [5,6], a specific hierarchical/modular modeling approach is adopted in order to better cope with system complexity and state-space explosion problems. [7] deals with the modelling and evaluation of phased-mission systems devoted to space applications, proposing a two level hierarchical method that allows to model such systems and to master the complexity of the analysis. Unfortunately, all these works and the others we are aware of are limited in their applicability and alleviate, but not completely solve, the complexity of the problem. Therefore, as a universal methodology for modeling and evaluating all types of complex systems does not exist, we define in this paper an ad-hoc methodology specifically tailored for the wireless system under analysis.

The rest of this paper is organized as follows. Section 2 presents the system context and the measures of interest. Section 3 introduces the solution technique adopted to perform the QoS analysis, and provides an overview of the models defined to represent the GPRS infrastructure and the behavior of the resource management techniques. Then, in Section 4 the numerical results of the simulation studies are presented and discussed. Conclusions are finally drawn in Section 5.

2   The System Context and QoS Indicators

We address a generic GPRS infrastructure, whose topology results in clusters of cells partially overlapping. To cope with congestion events, which may affect GPRS cells, e. g. due to a temporary lack of a number of traffic channels or to failures of their architectural components (as detailed in [2]), we assume that appropriate RMTs are applied. Instead of focusing on a specific RMT, we consider the class of RMTs which operate congestion alleviation by reducing the traffic of the congested cells, which is redirected to the neighbor partially overlapping cells. That is, a cell resizing is performed, and those users in the area no more covered by the resized cell are assigned to a neighbor cell covering the area where the users are located (if such an overlapping cell exists; in general, some users can be lost because of the black-spot phenomenon). This implies that the user population attached to such neighbor cells increases, thus affecting the QoS of such cells. Once the congestion is overcome, a re-switching process is operated to restore the initial user population. In order to analyze the effects of the traffic reconfiguration, we developed a methodology which is based on defining and separately solving sub-models capturing the behavior of those cells involved in the traffic reconfiguration applied through the RMT, that are the congested cell (called the *sending* cell) and a varying number of neighbor cells (called *receiving* cells). We call this set of cells a *congestion-effect* cluster. At a certain instant of time, a number of cells in the overall GPRS infrastructure could be experiencing a congestion event. Since, as just said, the effects of applying a RMT are local to each *congestion-effect* cluster, the analysis of the congestion impact can be carried on independently for the different *congestion-effect* clusters. Concerning a single *congestion-effect* cluster, three scenarios could be theoretically observed: i) a *sending* cell overlaps with N *receiving* cells and no such *receiving* cells overlap with any other *sending* cell; ii) a *sending* cell overlaps with N *receiving* cells and at least one of such *receiving* cells overlaps with another *sending* cell; iii) two or more overlapping *sending* cells are surrounded by N *receiving* cells (not all overlapping with all the *sending* cells).

In many cases the congestion of a cell lasts a short time (e.g. in case the partial outage is caused by a software error that can be fixed in a few minutes restarting the software); then, the probability of having multiple congested cells in a *congestion-effect* cluster is low and it would be reasonable to neglect the cases ii) and iii) above, and restrict to consider scenario i) only. Therefore, in the following we will refer to the *congestion-effect* cluster scenario depicted in Figure 1(a). Anyway, accounting for the other situations would not require changing the principles at the basis of our methodology and the steps it is composed of, but necessitates some extensions to the developed models (especially for the case ii) where a cell may contemporary receive users from multiple *sending* cells, while case iii) would be simply treated considering the set of overlapping *sending* cells as a single *sending* cell).

As mentioned, we do not concentrate on a specific resource management technique, but we consider the class of techniques that ultimately result in a cell

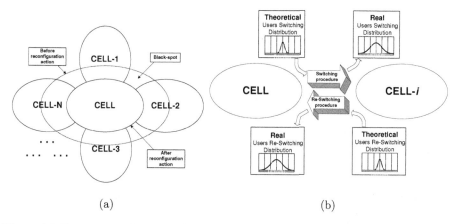

(a)   (b)

Fig. 1. (a) Congestion-effect Cluster and (b) Theoretical and Real Users Switching/Re-Switching Distributions

resizing or, equivalently, in a switching of users from one cell to another(s). The considered techniques are fully identified by the following characteristics:

1. the sending cell (CELL), that is the cell affected by outage;
2. the list of the receiving cells, that are the cells involved in the reconfiguration action (CELL-1, ..., CELL-N);
3. for each receiving cell CELL-i (with i=1, ..., N), the types of users to switch. A user may be: i) in the *idle* mode if he/she is not making any service request to the network system; ii) in the *active* mode if he/she is attempting to connect the network to get a service, and finally iii) in the *in-service* mode if he/she is connected and awaiting to get the service completed;
4. for each couple of cells [CELL,CELL-i] and type of users, the "theoretical users switching/re-switching distribution", that is the theoretical number of users that the technique expects to switch/re-switch at varying of time. It is only a theoretical distribution since, during the switching/re-switching phases, the number of available users can be lower than the corresponding theoretical value (Figure 1(b)), as we will emphasize later.

The goal of our analysis is to investigate the effects of outage, congestion treatment and outage recovery on the service provision, with special attention on the user perception of the QoS. More precisely, we aim to analyze the behavior of the network during the following temporal events (see Figure 2):

- At time T0, an outage occurs in the central cell (CELL), thus determining congestion some time after;
- At time T1, the switching procedure starts, causing some users to be switched from the congested cell to its adjacent ones;
- At time T2, the outage ends;
- At time T3, a Resource Management System (RMS) reacts to the end of the outage and starts the re-switching procedure from (CELL-1, ..., CELL-N) to CELL.

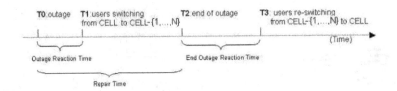

Fig. 2. Scheduled Temporal Events

We are interested in the following service availability measures:

- the point-wise congestion perceived by the users at varying of time (**PCf**), calculated as the *percentage of the unsatisfied users with respect to the total number of users in the cell.* An unsatisfied user is a user that is requiring a service but is not still served (active user);
- the total congestion indicator (**TCi**), inspired by [8], representing the *average congestion perceived by the users in a considered interval of time (E[PCf]).*

3   How to Model and Solve the System

The main problems in solving the model capturing the overall network's behavior are the time complexity (for the simulation) and the state space dimension (for the analytical solution), that rapidly increase if the number of receiving cells increases. Therefore, we investigated a modular approach, in which the solution of the entire model is constructed on the basis of the solutions of its individual sub-models. A simple, efficient solution would consist in splitting the overall model of Figure 1(a) in a number of simpler sub-models to be solved separately, for example one for each cell. In this case, the main problem we have to cope with is the temporal dependency between the congested cell and each of the receiving cells during the switching/re-switching procedure. In fact, as shown in Figure 1(b), the "theoretical" and the "real" users switching/re-switching distributions can be different, because of a lack of available users to be switched/re-switched at a specific time instant. For example, suppose that a RMT states to instantaneously switch X active users from CELL to CELL-i (theoretical distribution). If, at switching time, only Y active users are available (with $Y < X$), the switching procedure will follow a different (real) distribution: Y active users will be instantaneously switched, while $X - Y$ users will be switched one by one as soon as they become available.

To properly cope with this temporal dependency, we decomposed the overall model of Figure 1(a) in a set of more simple sub-models, each one composed by the couple [CELL,CELL-i]. The temporal dependency disappears as each sub-model manages the switching/re-switching procedure between sending and receiving cells.

In our developed methodology, a top-down approach is adopted to move from the entire system description to the definition of more simple sub-models.

Then, the model solution process follows a bottom-up approach: the solution of the entire model is constructed on the basis of the solutions of its individual sub-models.

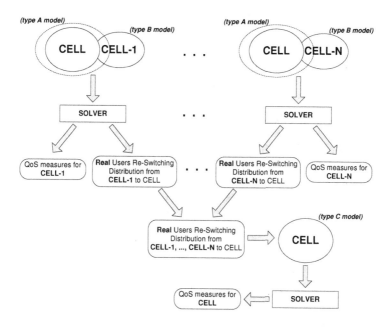

Fig. 3. Modeling and solution technique

It is a three step methodology. As it can be seen from Figure 3, we first decompose the overall model in N independent sub-models, each one composed by two cells: the first cell is always that affected by the outage (CELL), while the second is chosen from the other N receiving cells. Therefore, we solve N sub-models separately. From the solution of each single sub-model, we obtain two types of results for CELL-i:

- The QoS measures for CELL-i (a receiving cell), namely the percentage of unsatisfied users with respect to the total population;
- The "real users re-switching distribution", that is the real number of users re-switched from CELL-i to CELL as time elapses.

We note that in this first phase we do not obtain any information relevant to CELL, as each sub-model accounts for the re-switching procedure of only those users that have been previously switched from CELL to CELL-i, leaving out those users that have been previously switched from CELL to all other cells. In order to provide the QoS evaluations for CELL (the central cell), we perform another step in the solution technique. The "real users re-switching distributions" from each CELL-i to CELL are collected and combined, obtaining the "real users re-switching distribution" from CELL-1, ..., CELL-N to CELL. Finally, this distribution is given as input to another model (that represents the behavior of the central cell considering the re-switching procedure from all the neighbor cells to the central one) whose solution provides the QoS measures for CELL. We note that this last model requires the "real users re-switching distribution" as input, while the "real users switching distribution" is not explicitly required. This happens because we suppose that a receiving cell could not refuse an incoming user, and then the switching procedure only depends on the behavior of CELL (the sending cell).

3.1   The Types of Models Needed

In order to apply the methodology depicted in Figure 3 we need to construct three types of models only: type A, type B and type C. In this paper all the models are derived using Stochastic Activity Networks [9].

These models can be obtained as a specification of the model of Figure 4 representing an abstract view of a generic GPRS cell. The "internal GPRS cell model" was deeply described in [2], and it models the behavior of a GPRS cell during the random access procedure, when users compete to get a free channel. In fact, when a mobile station (MS) needs to transmit, it has to send a channel request to the network through the PRACH (Packet Random Access Channel), that is a channel dedicated to the uplink transmission of channel request. Since the network does not control the PRACH usage, the access method, based on a random access procedure, may cause collisions among requests by different MSs, and then may become a bottleneck of the system (see [10] for more details).

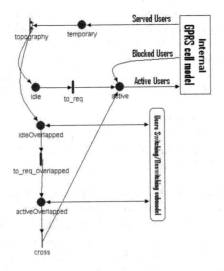

Fig. 4. A generic GPRS cell

The sub-model capturing the interactions between the central cell and the neighbor cells is the "users switching/reswitching sub-model". This sub-model has to be specified in order to:

- represent the behavior of the congested cell (CELL) during outage, cell resizing and outage recovery (type A model of Figure 3);
- represent the behavior of a receiving cell (CELL-i) during the resizing of the congested cell (type B model of Figure 3);
- represent the behavior of the congested cell (CELL) during outage, cell resizing and outage recovery using the provided "real users re-switching distribution" (type C model of Figure 3).

The generic model of Figure 4 works as it follows. When a user has been served, a token exits from the "internal GPRS cell model". This generic user has to be mapped (using the *topography* activity) in the overlapping area of the cell (place *idleOverlapped*) or in the non overlapping one (place *idle*), in accordance with the topography of the network. The probability that a generic user is mapped in the overlapping area is dynamically calculated considering the original number of users in the overlapping area and the overlapping users that have been switched to the other cells. When an idle user requests a new service, he/she becomes active and enters in the "internal GPRS cell model" that simulates the random access procedure of a GPRS cell. Finally, we note that the users switching and re-switching procedure affects only the users in the overlapping area, both in idle and in active mode.

For the sake of brevity we omit the definitions of type A and type C models (see [11]), while in the following subsections we present the model for the receiving cell CELL-i and the overall model for the couple of cells [CELL,CELL-i].

Type B model. Type B model represents the behavior of a receiving cell (CELL-i) during the resizing of the congested cell. It is obtained specifying the "users switching/reswitching sub-model" of Figure 4 as shown in Figure 5. The vertical black line separates the components belonging to the generic GPRS cell model (on the left) from those belonging to the "users switching/reswitching sub-model" (on the right).

Tokens in place *activeSwitched* (or *myActiveSwitched*) and *idleSwitched* (or *myIdleSwitched*) represent, respectively, the number of active and idle users really switched from CELL to CELL-i. The input gate *controller_switch_active* keeps the number of tokens in *activeSwitched* equal to the number of tokens in *myActiveSwitched*, until the re-switching procedure starts. The input gate *controller_switch_idle* performs the same action for the idle users. The *enable_reswitch* place contains one token if the re-switching procedure is enabled, zero otherwise. Tokens in places *commonActiveSwitched* and *commonIdleSwitched* represent, respectively, the active and idle users re-switched from CELL-i to CELL.

Here, we briefly describe the model behavior following the temporal events of Figure 2.

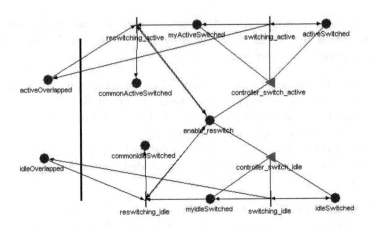

Fig. 5. "users switching/reswitching sub-model" for CELL-*i*

- Before time T0, the system is in steady-state.
- At time T1, the switching procedure from CELL to CELL-*i* starts and then some tokens arrive in places *activeSwitched* and/or *idleSwitched*. Places *myActiveSwitched* and *myIdleSwitched* follow the respective variations, thanks to the input gates *controller_switch_active* and *controller_switch_idle*.
- At time T2 the outage in CELL ends and then, at time T3, the mark of the place *enable_reswitch* is set to 1 and the re-switching procedure starts. The users re-switched from CELL-*i* to CELL are available in place *commonActiveSwitched* and *commonIdleSwitched*. The re-switching procedure ends when places *myActiveSwitched* and *myIdleSwitched* are empty.

Overall model for [CELL,CELL-*i*]. In the first step of the solution technique depicted in Figure 3, type A and type B models have to be composed together in order to build the overall model representing the behavior of each couple of cells [CELL,CELL-*i*]. The two models are joined together using the *Join*[1] operation [12] provided by the Möbius tool [13], and interact each other through the following shared places: *activeSwitched*, *idleSwitched*, *commonActiveSwitched*, *commonIdleSwitched*, *enable_reswitch*.

3.2   About Effectiveness

The major characteristic of this technique is its capability to manage the complexity of the overall model, as we provide the solutions solving N+1 sub-models only and combining some basic QoS measures. In case of state-based analytical solution, the state-space explosion problem is drastically reduced thanks to

---

[1] The Join operator takes as input a) a set of submodels and b) some shared places owning to different submodels of the former set. Its output is a new model that comprehends all the joined submodels' elements (places, arcs, activities) but with the shared places merged in a unique one.

the lower number of states generated for each individual sub-model. In case of simulation, the major advantages are related to:

- the mitigation of the stiffness-problem, if the submodels to be simulated during Step 1 and 3 have less time scales than the monolithic model. This property could be extremely useful in dealing with an heterogeneous network composed by cells of different technologies, e.g. GPRS and UMTS (Universal Mobile Telecommunications System);
- the decrement of the overall solution time, since the N sub-models constituted by the couple [CELL,CELL-i] in Step 1 can be solved concurrently. This favors the scalability of the method, which can easily deal with high numbers of receiving cells;
- the alleviation of the memory requirements for the simulator, as the sizes of the sub-models to be solved are reduced thanks to the models decomposition.

Although both analytical and simulation solution methods can be applied, in this paper we adopt the simulation approach to numerically solve the sub-models obtained applying our methodology, using the simulator offered by the Möbius tool. The main advantage in using the simulation is that it allows to represent real system conditions better than analytical approaches do (e.g., to use distribution functions more realistic than the exponential one).

4   Model Evaluation

We perform a transient analysis in the interval of time from the occurrence of an outage (time T0) to the new system steady-state after the outage repair.

4.1   Settings for the Numerical Evaluation and Analyzed Scenario

We analyze a GPRS network composed of one central cell (CELL) and three partially overlapping cells (CELL1, CELL2 and CELL3). In Figure 6 we detail the values we assigned to the main parameters of each cell. All the four cells have the same number of traffic channels (three) but different user populations; therefore, each cell has a different workload level at steady-state.

We analyzed two scenarios, which have been set up in order to tune the following two parameters of a resource management technique: *activeUsersToSwitch*, that is the number of active users to switch, and *outageReactionTime*, that is the time necessary to the Resource Management System to react to the outage.

- SCENARIO 1: The fine-tuning is performed in terms of the number of active users to switch from CELL to each other cell. In particular, we consider three cases: *i*) the case where no cell resizing is performed (no users switching), *ii*) the case where the cell resizing involves 50% of the users in the overlapping area (active users to switch = 75), and *iii*) the case where the cell resizing involves 100% of the users in the overlapping area (active users to switch = 150). Moreover, we set the *outageReactionTime* parameter to 30 seconds and assumed that 10% of the switched active users are lost during the reconfiguration action.

	CELL	
Users	180	
Overlapped Users	150	with **CELL1:** 60
		with **CELL2:** 50
		with **CELL3:** 40
Active Users to Switch	0, 75, 150	to **CELL1:** 0, 30, 60
		to **CELL2:** 0, 25, 50
		to **CELL3:** 0, 20, 40
Act. Users to Lose	0, 8, 15	to **CELL1:** 0, 3, 6
		to **CELL2:** 0, 3, 5
		to **CELL3:** 0, 2, 4

Users in *CELL1*	140
Users in *CELL2*	170
Users in *CELL3*	200

Outage Reaction Time	variable

Fig. 6. Analyzed scenario: cell topography and fine-tuning parameters

- SCENARIO 2: The number of active users to switch from CELL to the other cells is set to 75 users (30 to CELL1, 25 to CELL2 and 20 to CELL3). The focus in this scenario is on evaluating the impact of the time necessary to the Resource Management System to apply a traffic reconfiguration after the occurrence of an outage. So, the parameter under tuning is *outageReaction-Time*, for which three values have been considered: 15, 45 and 75 seconds. This performance indicator is useful to set a maximum value on the time the RMS is allowed to spend to elaborate a reaction to the observed overload.

We suppose that the switching and re-switching procedures are instantaneous. Moreover, we suppose that the partial outage affecting the central cell consists of a software error that reduces the number of available traffic channels from 3 to 1, and we set the outage duration to 120 seconds (average time needed to restart the software). The *outageEndReactionTime* parameter (the time that occurs between the end of the outage and the users re-switching) is set to 15 seconds (typical real value). In all the simulations we choose a relative confidence interval of 0.1 and a confidence level of 0.95, that is in the 95% of the times, the mean variable will be within 10% of the mean estimate.

4.2   Numerical Evaluation

In this section we show the results obtained from the simulations, both concerning the Pointwise Congestion function (PCf, on the Y-axes) and the Total Congestion indicator (TCi, in the labels of the figures). In all the figures plotting the simulation results, the time interval on the x-axis starts at time 200 sec. (the outage occurrence time) and ends at time 556 sec. (the time the new steady-state is reached in all the cells). The labels T0, T1, T2 and T3 on the x-axis have the same meanings as in Figure 2.

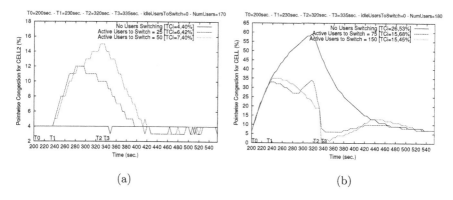

Fig. 7. (a) Congestion Perceived in CELL2 and (b) Congestion Perceived in CELL

Evaluation in scenario 1: Tuning of parameter 'activeUsersToSwitch'.
Figures 7(a) shows the congestion perceived by the users (the Point-wise Congestion function) in CELL2 at varying of the number of the active users to switch (0%, 50%, 100% of the number of users in the overlapping area). Obviously, the TCi value increases when we increase the value of the *activeUsersToSwitch* parameter. We note that the congestion level at steady state (time T0) is about 4%, after time T1 (the switching time), the congestion initially increases, but decreases immediately after. This happens when the receiving cell is not congested and, then, can absorb the added traffic. The other two receiving cells (CELL1 and CELL3) behave similarly and they are not presented in the paper for the sake of brevity. They only vary in the workload at steady-state level that is lower for CELL1 (about 1%) and higher for CELL3 (about 14%), mainly because of a different number of users camped in. Moreover, the traffic overload induced in CELL3 has the most negative impact, as the congestion level at steady-state is the highest.

Figure 7(b) shows the congestion perceived by the users in the cell affected by the outage at varying the number of the active users to switch from this cell to all the adjacent cells. From the figure we note that if we increase the total number of active users to switch from 75 to 150, the TCi value remains the same. This happens, in general, when the system tries to switch "too many" users and then the negative effects due, for example, to the augmented number of lost users is equivalent to the positive effects due to the augmented number of switched users. At time T1 the switching procedure starts and the perceived congestion is beneficially affected by the actuation of the technique in a very short amount of time. At time T2 the outage ends, CELL starts working properly and the congestion rapidly decreases, while increases from time T3 (because of the users re-switching), till reaching again the steady-state level.

Figure 8(a) shows the behavior of the overall GPRS network composed of CELL, CELL1, CELL2 and CELL3 at varying values of the *activeUsersToSwitch* parameter. We analyze the percentage of the unsatisfied users in the network with respect to the total number of users camped in the four cells (in this example

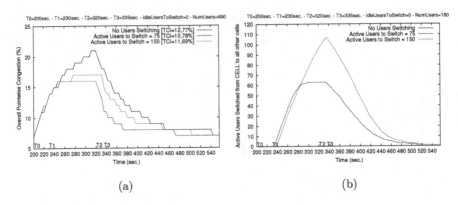

Fig. 8. (a) Overall Congestion Perceived and (b) Active Users Switched from CELL to all other cells

180+140+170+200=690 users). We note that the 100% cell resizing curve (*activeUsersToSwitch*=150) is worse than the 50% one (*activeUsersToSwitch*=75) as the positive effects induced by the decongestion in CELL don't compensate the negative effects on CELL1, CELL2 and CELL3 (the receiving cells).

Lastly, Figure 8(b) shows the number of active users really switched from CELL to the other cells. We note that the switching and re-switching procedures are not instantaneous. This means that there are not enough active users immediately available to be switched at time T1 (the switching time) and re-switched at time T3 (the re-switching time).

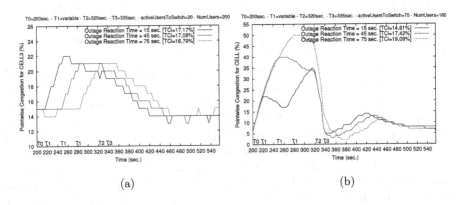

Fig. 9. (a) Congestion Perceived in CELL3 and (b) Congestion Perceived in CELL

Evaluation in scenario 2: Tuning of parameter 'outageReactionTime'.
Figure 9(a) shows the congestion perceived by CELL3 (one of the receiving cells) at varying the time needed by the system to react to the outage (*outageReaction-Time* parameter). As expected, the congestion increases if the outage reaction

time decreases, both concerning PCf and TCi, as the switched users reach the cell earlier. The other receiving cells behave similarly (they only vary in the workload at steady-state level) and then, for the sake of brevity, they are not presented in the paper.

Figure 9(b) shows the congestion perceived by the users camped in the central cell at varying of the *outageReactionTime* parameter. As expected, the TCi decreases when reducing the outage reaction time, as the reconfiguration action is applied earlier.

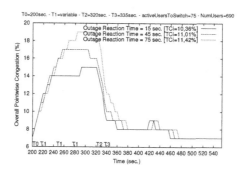

Fig. 10. Overall Congestion Perceived

Finally, Figure 10 shows the percentage of unsatisfied users in the overall network at varying the *outageReactionTime* parameter. This is the percentage of unsatisfied users in the network with respect to the total number of users camped in it (690 users for the considered setting). We note that if the reaction time parameter increases, the congestion perceived increases as well. The obtained results allow performing an interesting investigation on the amount of time that the system should be permitted to spend for its decision-making processes. For example, if a maximum tolerable level of degradation is known a priori, by looking at the results in Figure 10 it can be inferred a value for the maximum *outageReactionTime*.

5 Conclusions

In this paper, the congestion analysis of GPRS infrastructures consisting of a number of cells partially overlapping has been performed in terms of QoS indicators expressing a measure of the service availability perceived by users. When a congestion is experienced by one of these cells, a family of congestion management techniques is put in place, to operate a redistribution of a number of users in the congested cell to the neighbor ones, in accordance with the overlapping areas. Since the service availability perceived by users is heavily impacted by the congestion experienced by the cells, determining appropriate values for the users to switch, so as to obtain an effective balance between congestion alleviation in the congested cell and congestion inducement in the receiving cells, is a critical aspect in such contexts.

In order to carry on such fine-tuning activity, a modeling methodology, appropriate to deal with the system complexity, has been defined. In particular, we introduced a solution technique following a modular approach, in which the solution of the entire model is constructed on the basis of the solutions of the individual sub-models.

Models solution through a simulation approach has been performed in order to provide numerical estimates. The obtained results, although dependent on the considered parameters setting, show behavior trends very useful to make an appropriate choice of the number of users to switch, which is a critical parameter for the congestion management technique. Moreover, an investigation on the amount of time that the system should be permitted to spend for its decision-making processes is carried on.

The defined modeling framework shows very attractive potentialities, being it suitable to be employed in the analysis of other similar problems. Among the devised future works on this stream, we mention two directions. On one side, we could adapt this method to deal with other interesting scenarios, e.g. when a cell is overlapped with several congested cells. On another side, it could be reused to analyze the behavior of a heterogeneous infrastructure, where different network technologies (e.g., GPRS and UMTS) cooperate to reduce a congestion situation. This last is a direction we already started to explore in the context of the CAUTION++ project.

Acknowledgments

This work has been partially supported by the European Community through the IST-2001-38229 CAUTION++ project and by the Italian Ministry for University, Science and Technology Research (MURST), project "Strumenti, Ambienti e Applicazioni Innovative per la Societa' dell'Informazione, SOTTOPRO-GETTO 4". The authors want also to acknowledge the contribution given by Stefano Porcarelli to the early phases of this work.

References

1. IST-2001-38229 CAUTION++ Project. CApacity and network management platform for increased Utilization of wireless systems of next generaTION++. http://www.telecom.ece.ntua.gr/CautionPlus/
2. S. Porcarelli, F. Di Giandomenico, A. Bondavalli, M. Barbera, I. Mura. Service Level Availability Estimation of GPRS. *IEEE Transactions on Mobile Computing*, Vol. 2, N. 3, 2003.
3. P. Lollini, A. Bondavalli, F. Di Giandomenico, S. Porcarelli. Congestion Analysis during Outage, Congestion Treatment and Outage Recovery for simple GPRS networks. In Proc. of the Ninth *IEEE Symposium On Computers And Communications* (ISCC'2004), Alexandria, Egypt, June 28 - July 1, 2004.

4. D. D. Deavours and W. H. Sanders. An efficient disk-based tool for solving very large Markov models. *Performance Evaluation*, vol. 33, pp. 67-84, 1998.
5. N. Fota, M. Kaaniche, and K. Kanoun. Dependability Evaluation of an Air Traffic Control System. In Proc. Third *IEEE Int'l Computer Performance and Dependability Symp.* (IPDS), pp. 206-215, 1998.
6. K. Kanoun, M. Borrel, T. Moreteveille, and A. Peytavin. Availability of CAUTRA, A Subset of the French Air Traffic Control System. *IEEE Trans. Computers*, vol. 48, no 5, pp. 528-535, May 1999.
7. A. Bondavalli, I. Mura, M. Nelli. Analytical Modelling and Evaluation of Phased-Mission Systems for Space Applications. In Proc. of the *High-Assurance Systems Engineering Workshop*, Pages:85 - 91, 11-12 Aug. 1997.
8. Chang-Yu Wang; Logothetis, D.; Trivedi, K.S.; Viniotis, I.; Transient behavior of ATM networks under overloads. In Proc. of the *Fifteenth Annual Joint Conference of the IEEE Computer Societies. Networking the Next Generation* (INFOCOM '96), Page(s):978-985, vol.3, March 1996.
9. W. H. Sanders, and J. F. Meyer. A Unified Approach for Specifying Measures of Performance, Dependability and Performability. In Dependable Computing for Critical Applications, volume 4 of *Dependable Computing and Fault-Tolerant Systems*, pages 215-237. Springer Verlag, 1991.
10. ETSI, "Digital Cellular Telecommunication System (Phase 2+); General Packet Radio Service (GPRS); Mobile Station (MS)Base Station System (BSS) Interface; Radio Link Control/Medium Access Control (RLC/MAC) Protocol." GSM 04.60 version 8.3.0 Release 1999.
11. P. Lollini, F. Di Giandomenico, A. Bondavalli and S. Porcarelli. Congestion Analysis in a Multi-Cell GPRS Network. ISTI-CNR 2004-TR-26, http://dcl.isti.cnr.it/Documentation/Papers/Techreports.html
12. W. H. Sanders. "Construction and solution of performability models based on stochastic activity networks". Ph.D. dissertation, University of Michigan, 1988.
13. D. Daly, D. D. Deavours, J. M. Doyle, P. G.Webster, and W. H. Sanders. Möbius: An Extensible Tool for Performance and Dependability Modeling. In 11th International Conference, TOOLS 2000, volume *Lecture Notes in Computer Science*, pages 332-336, Schaumburg, IL, 2000. B. R. Haverkort, H. C. Bohnenkamp, and C. U. Smith (Eds.).

Characterizing Session Initiation Protocol (SIP) Network Performance and Reliability

Vijay K. Gurbani, Lalita J. Jagadeesan, and Veena B. Mendiratta

Bell Laboratories, Lucent Technologies
Naperville, Illinois
{vkg,lalita,veena}@lucent.com

Abstract. The Session Initiation Protocol (SIP) has emerged as the preferred Internet telephony signaling protocol for communications networks. In this capacity, it becomes increasingly essential to characterize both the performance and the reliability of the signaling entities utilizing the protocol. We provide an analytical look at the performance of a SIP network as well as a reliability model of SIP servers. Keywords: SIP, Stochastic Processes, Queueing Analysis, Performance Analysis, Reliability Analysis.

1 Introduction

The Public Switched Telephone Network (PSTN) has evolved over a century to become an integral part of human communications. Over the years, the network has been tuned for performance and has evolved to become highly reliable, with individual switches experiencing only a few seconds of downtime per year. As the telecommunications industry moves towards a new network (the Internet) with a new set of signaling protocols, media behaviors and routing protocols – which are markedly different from the PSTN model – is it reasonable to assume that the performance and reliability metrics established in the PSTN are applicable and achievable in the new environment?

Performance analysis and reliability of circuit-based communication networks has been well studied. Models exist in the PSTN that characterize performance in telecommunication switches. One measure of performance in the PSTN is the Busy Hour Call Attempt (BHCA) metric, which is defined as the number of call attempts during the busiest hour of the day. The BHCA measures the capacity of a PSTN call processing switch in terms of the total number of calls arriving at a switch during peak periods. In commercial PSTN switches, it ranges from 1 million to 2 million calls per hour. Another measure of performance is the switch cross-office delay, where a typical value is 100-300 milliseconds (ms); the precise requirements for this metric are specified by signaling message type.

Circuit switches for voice meet stringent requirements for reliability with expected switch availability greater than 0.99999 and expected call loss of the order of tens per million calls handled. For call loss, in the event of failures, the priority is to save calls in progress over calls in the setup stage. This high reliability is achieved through

M. Malek, E. Nette, and N. Suri (Eds.): ISAS 2005, LNCS 3694, pp. 196–211, 2005.

redundancy of the switch elements, robust software and the implementation of hardware and software fault tolerance mechanisms at various layers in the system.

Current trends in the telecommunications industry favor voice over Internet Protocol (VoIP) technology. The introduction of the Session Initiation Protocol (SIP) and the widespread adoption of the protocol by both wireless and wireline telecommunication players has accelerated the trend. If VoIP is to become the pervasive telecommunication model, then the performance and reliability of call processing elements in the Internet needs to be on par with those of the circuit-switched elements. To this end, there are three contributions of this paper. The first is to provide analytical models for the performance analysis of a SIP network and use the models to analyze the performance of a SIP network with respect to varying arrival rates, service rates and network delays. The network delay is characterized using one intermediary as well as a chain of intermediaries of varying length. The second contribution of the paper is evaluating a SIP network for reliability and lost calls. Given the industry trend towards using commercial-off-the-shelf hardware and software components, our evaluation is based on utilizing generally available application layer fault tolerance mechanisms as opposed to using proprietary solutions implemented at lower layers. Finally, we compare our findings with the established norms of PSTN performance and reliability.

The rest of the paper is organized as follows: Section II covers existing work related to SIP performance. Section III provides a brief background on the mechanisms of signaling exchange in the PSTN and a SIP network. Section IV details the performance model and the results from the performance analysis. Section V presents a reliability model combined with the performance model and the subsequent results. We conclude the paper by summarizing our observations and future work to be done in this area.

2   Related Work

Wu et al. [2] analyze SIP performance in light of SIP-T (SIP for Telephones) [3]. SIP-T is an effort to provide the integration of legacy telephone signaling into SIP messages through encapsulation and translation. The PSTN call setup messages that would normally flow between two PSTN switches are encapsulated and transported as a payload over a SIP network connecting two PSTN islands. SIP-T also translates certain PSTN call setup headers into their closest SIP equivalent to enable intermediaries to route the request. Wu et al. analyze the queuing delay and queuing delay variation using embedded Markov chains in a M/G/1 queuing model. Our work, by contrast, analyzes performance under varying arrival rates, service rates and network delays of an end-to-end native SIP ecosystem which includes multiple intermediaries (SIP proxies). We also analyze the reliability, including call loss, of SIP signaling entities through a hierarchical performance and reliability model.

The SIPStone benchmark [4] is an early attempt at characterizing server performance in a way that is useful for dimensioning and provisioning a SIP network. One of the aims of SIPStone is to enunciate a repeatable set of experiments in order to compare different implementations across the uniform set of experiments. It assumes the standard SIP trapezoid: a client conversing with a SIP intermediary, which in turns converses with a destination server. Our work builds in part on SIPStone to provide an analytical view of performance and reliability across a wider spectrum which includes modeling a SIP network using one intermediary, and a chain of intermediaries.

Zhu [10] analyzes the usage of SIP in the Third Generation Partnership Project's (3GPP) IP Multimedia Subsystem (IMS). This analysis involves the usage of SIP in the context of a centrally controlled architecture, which imposes additional requirements on the protocol above and beyond those specified in [1]. Our analysis is based on the protocol as specified in [1].

Lipson [12] presents an approach for using model checking of Markov Reward Models to analyze properties of a simple SIP network. The focus is on transient properties related to the number of calls processed prior to system failure or system repair. Rewards are expressed as simple rates of incoming requests for call setups. Our model, in contrast, is a hierarchical model consisting of a high-level Markov Reward Model and a lower-level queuing network model. Furthermore, our model considers implications of different fault tolerance approaches and we use closed-form equations rather than model checking to analyze properties of our model.

3 Background

In order to study the performance of telecommunications systems, it is instructive to understand the entities involved in call setup. We provide a brief overview of call setup in the PSTN and compare it with call setup in the Internet using SIP.

PSTN Call Setup. In the PSTN, telephone users connect through the telephone system into the central office (CO). Hundreds of COs may be installed in a metropolitan area. Telephone traffic from end users terminates at the CO through a pair of wires (or four wires) called the local loop or the subscriber loop. Telephone traffic from the COs is generally aggregated into trunks and carried to a toll/tandem office from where it is distributed to other toll offices. High usage trunks are established when the volume of calls warrants the installation of high capacity between two offices.

A salient point about the PSTN is that the network used to route the media stream between switches is different from the network used to route signaling messages. Signaling messages between switches are routed over a packet-based network called Signaling System Number 7 (SS7). Communicating switches exchange SS7 messages to setup a call by allocating media resource end-to-end. Once the media resources have been allocated and the call has been set up, the voice flows over direct media connections between each intervening switch. More information about PSTN signaling is available in [7].

Call Setup in SIP. SIP [1] is an application-layer protocol used to establish, maintain and tear down multimedia sessions. It is a text-based protocol with a request-response paradigm. A SIP ecosystem consists of user agents, proxy servers, redirect servers, and registrars. Of special interest to us with respect to this paper are user agents and proxy servers.

There are two types of SIP user agents: a user agent client (UAC) and a user agent server (UAS). A UAC and a UAS are software programs that execute on a computer, an

Internet phone, or a personal digital assistant (PDA). A UAC originates requests (i.e. start a phone call) and a UAS accepts and acts upon a request. UASes typically register themselves with a registrar, which binds their current Internet Protocol (IP) address to an email-like identifier used to identify the user. This registration information is used by SIP proxy servers to route the request to an appropriate UAS.

Proxy servers are SIP intermediaries that provide critical services such as routing, authentication, and forking. A SIP proxy, upon the receipt of an incoming call setup request, will determine how to best route the request to a downstream UAS.

The request to establish a session in SIP is called an INVITE. An INVITE request generates one or more responses. Responses to requests indicate success or failure, distinguished by a status code. Responses with status code 1xx (100-199) are termed provisional responses and serve to update the progress of the call; the 2xx code is for success and higher number for failures. 2xx-6xx responses are termed as final responses and serve to complete the INVITE request. The INVITE request is forwarded by a proxy (through possibly another chain of proxies) until it gets to its destination. The destination sends one or more provisional responses followed by exactly one final response. The responses traverse, in reverse order, over the same proxy chain as the request. Figure 1 provides a time-line of call establishment between a UAC and a UAS. The request is forwarded through a chain of proxies.

With reference to Figure 1, the UAC sends an INVITE to P1 and P1 routes the call further downstream. From the UAC's reference, P1 is called an outbound proxy. P1 determined that the request should be forwarded to P2 (the UAS is in a different domain). When the request arrives at P2, it queries its location server and further proxies the request to the UAS. From the UAS point of view, P2 is the inbound proxy. The UAS issues a provisional response followed by a final response. The call is setup when the UAC receives the final response.

Comparing SIP entities to the PSTN, the UAS and UAC correspond to phones; proxies act as 'switches'. However, unlike the PSTN, there is no signaling overlay network. Both media and call signaling use the same network. Nor is there a notion of a toll/tandem switch in the Internet. The routing fabric of the Internet assures that packets containing voice or data are forwarded to their intended destination. More information on Internet telephony signaling and SIP is available in [1, 8, 9].

4 Performance Analysis

The performance measures of interest for SIP networks are the steady-state mean response time and mean number of jobs in system. The mean response time of a proxy server is defined as the mean elapsed time from the time t_1 an INVITE request from an User Agent Client (UAC) arrives at the proxy server until the time t_2 that the proxy server sends a final response to the UAC. The mean number of jobs in system is defined as the mean number of calls currently being set up or waiting to be set up by the proxy server. Also of interest is the behavior of these performance measures as a function of the mean arrival rate of incoming INVITE requests, the mean service rates for processing SIP requests/responses, and the mean propagation delay between adjacent SIP proxy servers in the network.

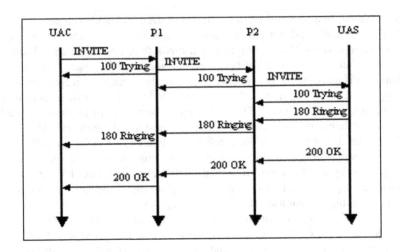

Fig. 1. SIP call establishment

4.1 Performance Model and Assumptions

We model a SIP proxy server as an open feed-forward queuing network, in which arriving jobs correspond to INVITE requests received by the SIP proxy server from an upstream UAC in a SIP network. The queuing network consists of sequences of queuing stations that correspond to possible sequences of SIP requests and responses during a call setup.   Each queuing station does the servicing of the SIP request/response at the corresponding point in the call setup sequence. In constructing our model, we made some simplifying assumptions. First we model a "180 Ringing" response and assume that immediately following this will be a final response (either a 2xx final response or non-2xx final response). When an INVITE request arrives at the proxy, it is sent downstream and may engender a "180 Ringing" response or a non-2xx final response.

Next, we make certain assumptions about the mean service time.   In SIP, mean service time will vary by implementation. For this analysis, we assume that it takes $1/\mu$ mean time to service an INVITE request at a proxy and derive other service time parameters from this base service time. Servicing a SIP message includes extracting the message from the transport layer, parsing it, performing a location server lookup, querying the DNS and serializing the request on a connection opened with the next downstream entity. In response to the INVITE, the proxy will receive a 180, a 200, or a non-200 response. Since the effort required to process a response is far less than that for processing an INVITE, we assign a mean service time of $0.3/\mu$ for processing 180, 2xx and non-2xx responses.

For simplicity, we assume a lossless network. This is not an unreasonable assumption, loss rates of 10^{-7} are not uncommon in Internet2 [13].   Operational networks will typically have very low packet loss rates to maintain good voice quality and acceptable call setup delays. Finally, we assume a simple call flow from a UAC to an outbound proxy, which transmits the call to an inbound proxy in the domain of the UAS and from there it arrives at the UAS. For this call flow, we model two cases: one, the inbound proxy is the same as the outbound proxy (UAC $\Rightarrow$ P $\Rightarrow$ UAS), and, two, there is a chain of proxies between the UAC and the UAS (UAC $\Rightarrow$ P$_1$ $\Rightarrow$ P$_2$ $\Rightarrow$ ... $\Rightarrow$ P$_N$ $\Rightarrow$ UAS). We do not consider advanced SIP services such as forking. Figure 2 shows the basic model.

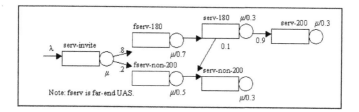

Fig. 2. Model with no network delays

When an INVITE arrives at a proxy, with a probability of 0.8 it will engender a "180 Ringing" response, and with a probability of 0.2 it will result in a failure response. The failure leg models the behavior of a call that was not setup. Following the model further, we note that with a probability of 0.9, the "180 Ringing" is followed by a "200 OK" response; i.e. the user associated with the UAS successfully answered the call. With a probability of 0.1, we model the "180 Ringing" resulting in a non-2xx response; i.e. the UAS was successfully contacted but the user did not pick up the phone. The fserv-180 and fserv-non-200 stations model a UAS. A UAS does not proxy a request downstream; instead, it issues a response. As such, it requires less computation than what a proxy undergoes when it services a request. Hence, in the model, we have assumed a mean service time of $0.7/\mu$ for sending the 180 followed by a 200 or non-2xx response. Similarly, sending only a non-2xx response takes even less time, modeled by a mean service time of $0.5/\mu$. Note that we assume that there is zero delay between the "180 Ringing" response and the "200 OK" response. In real systems there will be a variable delay — this is the time taken by the user to answer the call. The length of this delay interval would impact the number of jobs in system performance measure and also has implications for checkpointing.

In the base queueing model of a SIP proxy server depicted in Figure 2, each queueing station is modeled as a M/M/1 queue. This model is an open, feed-forward queueing network, since jobs arrive from an outside source, and there is no feedback among queueing stations in the queueing network. Using standard approaches [11], the mean number of jobs N in system is given by $N = \sum_{k=1}^{J} \rho_k/(1 - \rho_k)$, where $\rho_k = \lambda_k/\mu_k$, $\lambda_1 = \lambda$, $\lambda_j = \sum_{k=1}^{j-1} (\lambda_k Q[k,j])$ for $1 < j \le J$, and J=6 is the number of stations in the queuing model. Q is the one-step probability matrix corresponding to the queuing model, that is, Q[i,j] is the probability that a job departing station i goes to station j. Since the queuing network is feed-forward, we assume that the serv-INVITE station corresponds to station 1, and the other stations are numerically ordered in the above equations so that Q[i,j] = 0 for all i ≥ j. The mean response time R for jobs is then given by Little's law [11], R = N/λ.

We now extend this model to include propagation delays between adjacent SIP proxy servers and UAC and UAS's in call setup paths. Propagation delays can be modeled through a delay server; namely, a M/M/∞ queuing station with mean service time given by the mean propagation delay. The extended model is shown in Figure 3.

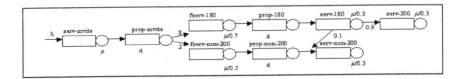

Fig. 3. Model with network delay

The prop-INVITE station models the propagation delay in proxying the INVITE request to the downstream SIP entity, while the prop-180 station models the propagation delay in receiving a 180 response, together with a 200 response or non-2xx response, from the downstream SIP entity. The prop-non-200 station is similar.

The mean number of jobs in M/M/∞ stations is given by the arrival rate of jobs into the station multiplied by the mean service time (i.e. mean propagation delay in our model) [11]. It is thus straightforward to extend the earlier equations to compute mean response time and mean number of jobs in system for this extended model. Note that the model in Figure 2 corresponds to the extended model with propagation delay of zero.

4.2   Results of Performance Analysis for SIP Proxy Servers

Using this approach, the mean response time for a proxy server is computed; the results are shown in Figure 4. The plots show propagation delays from 0 to 10 ms, corresponding to distances of 0 to 1000 miles between adjacent SIP entities assuming delays of 1 ms per 100 miles. The INVITE service rate is fixed at 0.5 ms[-1]. We observe that the mean response time is essentially linear with the arrival rate for the range of values considered. As expected, the mean response time increases with the mean propagation delay time. In our evaluated interval of arrival rates and propagation delays, the mean response time is in an acceptable range (as compared to the 100-300 ms for PSTN switches). Figure 5 shows the mean number of jobs in system as the arrival rate varies. We observe that the mean number of jobs is quite small (less than 10), even under propagation delays corresponding to a distance of 1000 miles.

We then compute these same measures of interest, this time varying the service rates for processing INVITE requests. Figures 6 and 7 show the mean response time and mean number of jobs in system as the service rate is varied. In this analysis, the arrival rate of INVITEs is fixed at 0.3 ms^{-1}; i.e. 1 million BHCA.

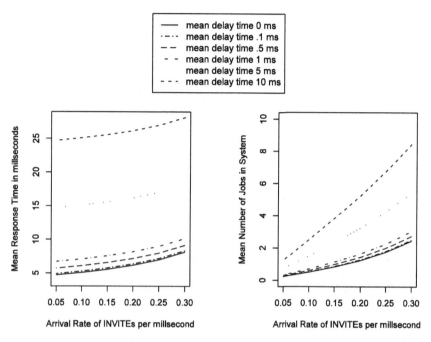

Fig. 4. Mean response time under varying arrival rates

Fig. 5. Mean number of jobs under varying arrival rates

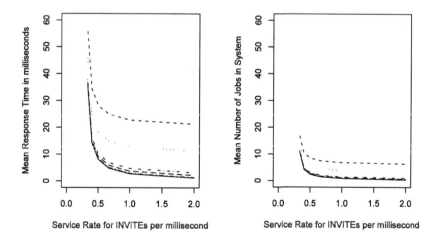

Fig. 6. Mean response time under varying service rates

Fig. 7. Mean number of jobs under varying service rates

4.3 Performance Model for Multiple SIP Servers

We now extend the model and analysis in two ways: first, to hosts running multiple proxy servers for scalability, and second, to a network of SIP proxy servers.

Multiple Proxy Servers on a Single Host

Clearly a single server solution for a proxy is not scalable. We therefore provide performance results for a multi-server proxy host. We extend the model of Figure 2 to queuing networks with the same structure, but with each M/M/1 queue replaced by a M/M/m queue. The equations for computing the mean response time and mean jobs in system are standard (c.f. [11]). Figure 8 depicts the performance results for the model of Figure 2 with M/M/m queues, where the number of servers, m is varied between 3 and 10, and the propagation delay is set to zero. The lower bound of 3 servers corresponds to the minimum number of servers needed to ensure that the queuing network is stable. A key observation from Figure 8 is that below a certain threshold for the service rate μ (i.e. 0.3 INVITEs ms^{-1}), the mean response time to process requests can grow significantly even under small changes in the service rate. Thus, this indicates the minimum service rate for multiple server hosts to ensure robustness of the proxy server. Our second observation is that for values of μ greater than this threshold, not only is the mean response time less sensitive to changes in the service rate, it is also largely independent of the number of servers in a single proxy server host. This implies that a small number of multiple servers with a service rate of 0.3 is sufficient, so large numbers of servers or faster service rates are not necessarily needed. Figure 9 depicts a similar analysis, where the mean network delay is fixed at 1 ms. The results are similar to Figure 8, with an increase in mean response time corresponding to the network delay.

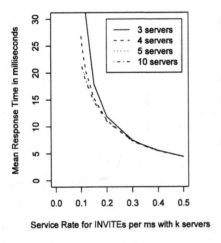 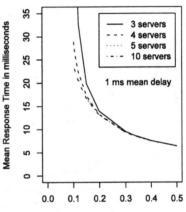

Fig. 8. Mean response time of multiple server host under varying service rates

Fig. 9. Mean response time of multiple server host with varying service rates and network delay

Chain of SIP Proxy Servers

We next extend our analysis of a single server host in an orthogonal direction: namely, to a network of proxy servers modeling multiple hops in an end-to-end network. We thus extend our performance measures of interest of mean response time and mean jobs in system to reflect the end-to-end network. In particular, the mean end-to-end response time is defined as the mean elapsed time from the time t_1 an INVITE request from an User Agent Client (UAC) arrives at the proxy server until the time t_2 that the proxy server sends a final response to the UAC; this mean response time now includes the time taken by all the intermediate proxies and the far end UAS to set up the call. Similarly, the mean number of jobs in system is now defined as the mean number of calls being set up or waiting to be set up by any of the intermediate proxy servers involved in setting up the call.

In order to do this analysis, we need to recursively replace each station modeling the far end in our queuing network by a copy of the queuing network. However, separately replacing the fserv-180 and fserv-non-200 stations by copies of the queuing network is incorrect, since the arrival rate into the copies of the queuing network would recursively be a fraction (0.8 or 0.2) of the arrival rate into the base model. Hence, this recursive model would incorrectly assume greater capacity in the system. We thus first use an alternative model to our queuing network in which the fserv-180 and fserv-non-200 stations are replaced by a single fserv station. This model is depicted in Figure 10, where Nserv is the sum of the mean number of jobs at stations fserv-180 and fserv-non-200 computed from the model of Figure 2.

It is straightforward to show that, for any arrival rate λ, if the service rate μ_{fserv} is given as $\lambda(Nserv+1)/Nserv$, the mean response time and mean number of jobs of this alternative model and the original model are equivalent. We thus construct our model of SIP networks by recursive substitution into this alternate model. In particular, we

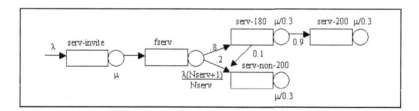

Fig. 10. Equivalent model for analysis

recursively substitute the fserv station in this alternate model with a copy of this alternate queuing network, and then compute the mean end-to-end response time and mean number of jobs in the end-to-end system. A similar construction is done for the extended model that included propagation delays. Figures 11 and 12 show the results of this analysis, where the length of the proxy chain is varied from 1 to 6. The different lines again correspond to varying the propagation delays. The arrival rate is fixed at 0.3 ms^{-1} and the service rate for INVITEs is fixed at 0.5 ms^{-1}.

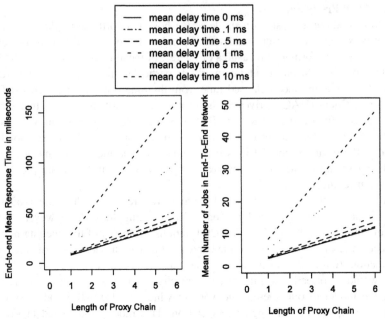

Fig. 11. End-to-end mean response time under varying length of proxy chains

Fig. 12. End-to-end mean number of jobs under varying length of proxy chains

5   Reliability Analysis

The reliability metrics of interest for the proxy server are the steady-state system availability and the probability of job loss (i.e. loss of SIP call requests). We first develop a standard reliability model for the single proxy server for various replications schemes. The reliability model is then combined with the queueing performance model of Section IV to predict the probability of job loss for these replication schemes. For this analysis, we used the hierarchical reliability and performability models and associated closed form expressions for computing availability and loss probability presented in [5].

The existence of fault tolerance software running at the application layer, that provides process and node error detection, recovery and checkpointing capabilities (as appropriate), is assumed for the proxy server. As in [5], the server is assumed to exhibit fail-silent behavior. When there is a server failure, the messages at the server, could be lost or saved depending on the recovery mechanisms implemented. The same applies to new messages arriving at the server during the detection and recovery intervals. Since the queuing network performance model assumes an infinite size buffer for the wait queue, messages are not lost due to buffer overflow. Thus, in the event of a server failure, the following 3 message loss scenarios are of interest: queued jobs, in-service jobs and new job arrivals when the system is down are lost (Case V in [5]); queued jobs, in-service jobs and new job arrivals when the system is down are not lost (Case II in [5]); and queued jobs and in-service jobs are lost and new jobs arrivals when the system is down are not lost (Case VI in [5]).

5.1 Reliability Models

Continuous Time Markov Chain (CTMC) models, which capture the failure, error detection and recovery behavior of the server are evaluated for the following replication schemes: no replication, cold replication and warm replication. Server failures are caused by process or node failures, and it is assumed that there is only a single failure in the system at any time.

No Replication. There is a single proxy server with no fault tolerance software. Error detection and recovery are done manually. The unavailability of the server is observed only after the failure is detected and recovery is initiated after detection.

Cold Replication. There are two proxy servers running in active/cold standby mode with fault tolerance software at the application layer. Upon detection of a failure of a process in the active node, the process is restarted and the system is returned to a working state; with some probability this may require switchover to the standby node. Upon detection of a failure of the active node the recovery action is to switchover to the standby node. In this case the switchover time includes the time required to bring up the node. We follow the cold replication model given in [5].

Warm Replication. There are two proxy servers running in active/warm standby mode with fault tolerance software at the application layer. In the event of active process or processor failure, the standby node assumes the role of the active node after detection of the failure and switchover. A new backup is started on another available node. In the event of standby process failure, the process is restarted or, if it exceeds the threshold of restarts, it is started on a different node. We follow the warm replication model given in [5].

For all of the above replication schemes, availability is calculated from the pure reliability models by adding the steady state probabilities of the server up states. The pure reliability models at the high level and the queuing models of Section IV at the lower level are combined to compute the probability of call loss. In particular, rewards are associated with each state and transition of the reliability model. Rewards associated with a state reflect the rate of expected loss of call requests in that state; lost call requests accumulate at the specified reward rate during the expected time spent in the state. Impulse-based rewards associated with transitions reflect the number of calls lost when the transition takes place. Expected rate of loss is computed by the accumulation of lost call requests in states and during transitions; we use the closed form equations from [5]. As in [5], loss probability is calculated by dividing the expected rate of loss of incoming jobs by the expected job arrival rate.

5.2   Results of Reliability and Call Loss

The following parameter values (with exponential distributions) are assumed for the reliability and call loss analysis of SIP proxy servers:

Job arrival rate, $\lambda = 0.3$ ms^{-1} Process failure detection rate $\delta_p = 1$ sec^{-1}
Job service rate, $u = 0.5$ ms^{-1} Manual recovery rate, $\tau = 1/120$ sec^{-1}
Process failure rate, $\gamma_p = 0.1$ day^{-1} Process restart rate, $\tau_p = 1/30$ sec^{-1}
Node failure rate, $\gamma_n = 0.05$ day^{-1} Process failover rate, $\tau_n = 1/120$ sec^{-1}

The node failure detection rate, δ_n is varied from 0.1 sec^{-1} to 15 sec^{-1}. The node switchover rate, τ_s, from failed to warm standby, is varied from 1/5 sec^{-1} to 1/30 sec^{-1}. The job (INVITE) arrival rate and service rate are as in the models of Section IV.

Figure 13 shows the availability of a proxy server for different values of the node failure detection rate for the case of no replication, cold replication and warm replication. Node availability greater than 0.9999 is achievable with warm replication and it is not sensitive to increases in the detection rate beyond 1 sec^{-1}.

Figure 14 shows the probability of job loss for a proxy server for different values of the node failure detection rate for no replication, cold replication and warm replication. The loss scenario is that queued jobs, in-service jobs and new job arrivals when the system is down are lost; therefore, no checkpointing is required. For all cases, there is an initial decrease in the probability of job loss as the node failure detection rate is increased from 0.1 sec^{-1} to 1 sec^{-1} and, for further increases in the detection rate, there is an increase in the probability of job loss except for the no replication case where it remains constant. The probability of message loss increases significantly for values of the detection rate greater than 12 sec^{-1} due to the increased overhead associated with the higher detection rate.

We assume that the buffer for job arrivals entering the system is of infinite size and, therefore, no jobs are lost due to buffer overflow. In Figure 15, we plot the expected number of job arrivals when the system is in the down state against the detection rate of node failures for different replication schemes. As expected, this figure is highest for the no replication case (longest downtime) and lowest for the warm replication case (shortest downtime). The results, however, are not sensitive to changes in the node failure detection rate beyond 1 sec^{-1}. Next, in Figure 16, we show the mean time required to service the jobs accumulated in the arrival queue (while the system was in a down state) as a function of the node failure detection rate. The service time for these jobs ranges from 75 seconds for no replication, 40 seconds for cold replication and 2 to 20 seconds for warm replication depending on the node switchover rate. The point to note is that, when the job arrival rate is high, saving job arrivals during the recovery interval is not worthwhile in call processing applications — the call setup delays for no replication and cold replication schemes would be unacceptable. This implies that checkpointing will not provide any benefits for the no replication and cold replication schemes.

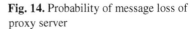

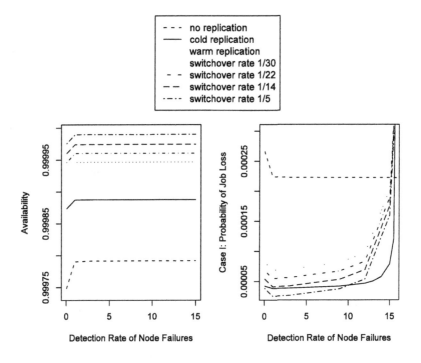

Fig. 13. Availability of proxy server

Fig. 14. Probability of message loss of proxy server

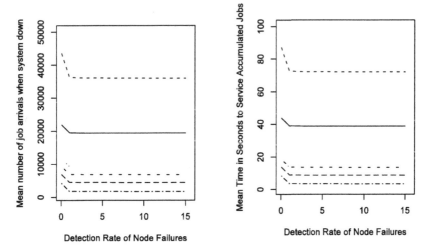

Fig. 15. Arrival of INVITE requests during down state

Fig. 16. Mean time to service jobs in arrival queue

6 Conclusions and Future Work

We have presented performance and reliability models for SIP networks and analyzed the behavior of the network under varying arrival rates, service rates, network delays, and replication schemes and associated failover rates. Key metrics that were analyzed include (end-to-end) mean response times, (end-to-end) mean number of jobs in the system, availability, probability of job loss, and mean time to process jobs that arrive when the system is down. Our analysis indicates three key findings. First, for an arrival rate of 1 million BHCA our results show the mean response time falls within an acceptable range, and that beyond a certain point, increases in service rates or number of servers on a single host do not yield significant improvements in mean response time. In particular, our results show that for single server hosts and service rates of 0.5 INVITE ms^{-1}, mean response times are less than 10ms. Furthermore, service rates greater than 1 ms^{-1} do not yield significant improvements in mean response time. Similarly, for multiple server hosts and service rates of 0.3 ms^{-1}, response times remain acceptable. Second, our results indicate that in steady state there are very few jobs in the system that are in a setup state. For example, in the steady state we observe that single server hosts with service rates greater than 0.5 requests per ms, there are no more than 10 jobs in the setup state in a single proxy server. For chains of single server proxies up to length 6, there are no more than 50 jobs in the setup state across all proxies in a SIP network. Given these results, we question whether it is necessary to add checkpointing in a SIP network. As noted earlier, however, if the delay representing the time taken by the user to answer the call is included in the analysis there will be more jobs in the system in a *ringing state*. Our future work will extend the performance analysis to multiple servers on hosts.

Third, our results demonstrate that saving incoming jobs when the system is down yields acceptable mean response times only under certain replication schemes. For no replication and cold replication, the mean time to service the INVITE requests accumulated during the recovery interval will require 40-75 seconds. Given that the normal lifetime of a SIP transaction is 32 seconds [1], saving job arrivals during the recovery interval is counter-intuitive. For warm replication, however, the mean time to service the jobs accumulated during the recovery interval is 2-20 seconds, depending on the value of the node switchover rate. For this replication strategy, one can consider saving new jobs that arrive during the recovery interval. However, since, as discussed above, calls in the setup state likely need not be saved, a comprehensive checkpointing strategy is not necessary. Our future work will also extend this aspect of our analysis with multiple servers on hosts.

The reliability model presented results for an assumed set of input parameter values. The results indicate that, to achieve the level of reliability in SIP networks that is comparable to PSTN, warm replication is required. In practice, these models can be used to determine required design targets such as switchover time and error detection time to achieve a given level of proxy server availability.

Future work will focus on validating the performance and reliability model parameters and results with lab measurements and field data. Additional future work includes relaxing assumptions about exponential distributions, including protocol timers in the model and extending the reliability model to multiple servers.

References

[1] J. Rosenberg, et al., "The Session Initiation Protocol (SIP)", *IETF RFC 3261*, June 2002, <http://www.ietf.org/rfc/rfc3261.txt>.

[2] J-S. Wu and P-Y Wang, "The performance analysis of SIP-T signaling system in carrier class VoIP network", Proceedings of the 17th IEEE International Conference on Advanced Information Networking and Applications (AINA), 2003.

[3] Vemuri and J. Peterson, "Session Initiation Protocol for Telephones (SIP-T): Context and Architectures", IETF RFC 3372, September 2002, <http://www.ietf.org/rfc/rfc3372.txt>

[4] H. Schulzrinne, et al., "SIPStone - Benchmarking SIP server performance", April 2002, <http://www.sipstone.org/files/sipstone_0402.pdf>.

[5] S. Garg, et al., "Performance and Reliability Evaluation of Passive Replication Schemes in Application Level Fault Tolerance," Proceedings of the 29th Annual International Symposium on Fault-Tolerant Computing, Madison, WI, June 1999.

[6] J. F. Meyer, "On evaluating the performability of degradable computing systems", IEEE Transaction on Computers, Volume 29, No. 8, pp. 720-731, August 1980.

[7] T. Russell, "Signaling System #7", (Second Edition), McGraw-Hill Publishing Company, 1995.

[8] G. Camarillo, "SIP Demysitified", McGraw-Hill Publishing Company, 2001.

[9] J. Davidson, et al., "Voice over IP fundamentals", Cisco Press, 2000.

[10] Zhu, "Analysis of SIP in UMTS IP Multimedia Subsystem", MSc. Thesis, Computer Engineering, North Carolina State University, 2003.

[11] R. Jain, "The Art of Computer Systems Performance Analysis", John Wiley and Sons, Inc., 1991.

[12] F. Lipson, "Verification of Service Level Agreements with Markov Reward Models," South African Telecommunications Networks and Applications Conference, September 2003.

[13] P. Barford and J. Sommers, "Comparing Probe- and Router-Based Packet-Loss Measurements," IEEE Internet Computing, Vol. 8, No. 5, pp. 50-56, September-October 2004.

Author Index

Lecture Notes in Computer Science

For information about Vols. 1–3615

please contact your bookseller or Springer